黃淑儀

給入廚新手的菜式

Must-Learn Recipes For Novice Chefs

Preface

寫給黃淑儀的新書

我愛吃。吃過和讀過的美食大半記得，老了還記得。主編《大成》雜誌的沈葦窗先生精醫道，精饌飲，早年和沈先生聊天最有趣，故事多得不得了：張大千親自下廚為清真回教徒馬連良做了一道雞肉獅子頭；李麗華愛吃鮑魚，家居飯菜必有鮑魚，慈母張少泉女士親自採購上佳貨色；張大千家的清燉鯽魚羊肚湯但求一個「鮮」字，因為「鮮」字一半是魚，一半是羊；大千説北方館子燒豬頭酥爛的秘訣是用普洱茶先煮豬頭，煮爛了棄掉茶渣加醋再煮，豬頭自必碎嫩如豆腐；徐悲鴻最愛鹹帶魚下飯；章太炎愛吃花生米；魯迅説廣東水果中楊桃最好吃，滑而脆，酸而甜；蘇州女人愛吃松子做的糖果，説話聲音格外柔媚，吳儂軟語也。沈先生這些故事後來都寫進他的名著《食德新譜》。

我結識黃淑儀數十年了，每次聽她説烹飪總是想起沈先生，總是慨嘆學問之大，不輸七藝。她做的菜不論盛宴不論家常，發明既多，驚喜更多，絕不討巧，也不花哨，樣樣好吃。梁實秋《雅舍談吃》自序説，一位先生問梁先生：「您為什麼對飲食特別有研究？」梁先生聽了惶恐，説他根本沒有研究過：「只因我連續吃了八十多年，沒間斷！」這句話説出境界最高是尋常，做菜的人不矯情，吃菜的人也不矯情，尋常之中見道行，那正是黃淑儀讓人敬愛的本領。陸放翁《老學庵筆記》記北宋年間開封炒栗子炒得最好的李和，説別的店舖怎麼仿效都比不上他。清代文豪袁枚的《隨園食單》連鹹蛋怎麼吃才好吃都寫了。這是美食的意趣更是廚藝的神髓，黃淑儀最懂。我討厭造作的廚師也討厭矯揉的菜餚，那跟亂擺山林架子的假名士一樣低俗。黃淑儀倒是鄰家會煮菜的媽媽，帶着鄉土情懷撫慰現代味蕾的能手。她和徐景清一家是我的舊鄰居，日落時分他們家炒菜的香氣裊裊飄散，至今不忘。夜雨剪春韭，新炊間黃粱，美好的老歲月。

董橋

二〇一九年己亥端午節在香島

我的第十四本 Cookbook！

去年接到公司指令，「吾淑吾食」要做「溫哥華篇」，當時的感覺是既興奮又擔憂！興奮者，身為烹調愛好者、飲食節目主持人，可以在生活了三十多年的溫哥華，名正言順的發掘她的食材，深入探討她的飲食文化，意義重大！擔憂者，面對眾多食材，如何能把她最好的一面傳揚出去，馬上要寫更多食譜，找更多前輩及加倍努力的鑽研，我能嗎？

誠惶誠恐！

在找到資料後，配合食材、配合要求，挖空心思的嘗試；説到底，在我的食譜裏，要的是簡單易做，而又要不失禮的家常菜，才能吸引觀眾（讀者）的興趣。

在烈日當空下的田畦、花園、露台、海邊、果園，都搭上爐頭煮番兩味。當中，除了一些改良舊菜譜之外，也有不少新菜式；例如，為了迎合中國人口味，把蘋果湯改頭換面，但卻不失為一道別緻的湯水。

辛苦了二十八日，終於完成了十三集溫哥華篇，完成了二十多個菜式、十三個湯，完成了個人的創舉！

不過三個月下來（連拍攝前期工作），我整個人都垮了！除了膝痛難當，要以拐杖輔助走路，水腫及失眠，身體不瘦反胖，我徹底認輸；也讓我深深體會到身體健康的重要！

老公訓示：不可再操勞，不可再接工作；要在家靜下心來，好好休養……！

也好！讓我有時間整理新食譜；於是，就有了這本糅合新節目與新嘗試的新 Cookbook！

不過，我這頭靜不下來的驢子，又接了新工作，要去台灣拍民宿，希望此行會給我帶來更多資訊及更多烹飪資料與大家分享。

邀請董橋先生為我的新書寫序，原以舊鄰居身份，向董夫人旁敲側擊，不敢直接問，怕被指高攀，怕被拒！到尾一句～～怕瘀！

認識董先生多年，每次見面都笑容可掬，雖然極少説話，但一旦發言，多有獨特見解。董先生學識淵博，著作頗多，其散文更獲獎無數！

* 消息傳來，董先生剛剛獲選為 2019 年「世界華文文學獎」得主，實在可喜可賀！

黃淑儀

2019 年 6 月 26 日

目錄 Contents

主菜

甜品 • 小吃

麵食

湯羹

醬

處理方法

#main dish

#sweet & snack

#noodle

#soup

#sauce & spread

#preparing tips

花旗參珍珠丸

Glutinous Rice Pork Balls with American Ginseng

材料

半肥瘦豬肉	半斤（切片，洗淨，瀝乾水後再剁碎。）
花旗參	2 克（切片，用適量水浸軟、剁碎，水留用。）
馬蹄	4 粒（去皮，放入膠袋內，用刀拍扁再切碎。）
糯米	1 杯（洗淨，浸過夜。）
杞子	適量（略洗）

做法

1. 免治豬肉加入鹽和糖各半茶匙、生粉 1 茶匙及花旗參水醃一會，拌勻後加入花旗參碎及馬蹄碎拌勻，擠成肉丸子。
2. 把肉丸子放在糯米上翻滾，讓每顆肉丸子都沾滿糯米，再飾上一顆杞子。
3. 用猛火蒸 10 分鐘即可。

GIGI NOTES

千萬不要加入葱，否則會搶去花旗參的味道。
將豬肉切片後再洗，可去除豬羶味。
糯米宜浸透才蒸，否則會不夠軟糯。

炸醬

Zha Jiang Pork Sauce

材料

梅頭	1 斤半（切絲）
乾葱頭	8 粒（切絲）
油	2 湯匙

調味料

茄汁	8 湯匙
紅辣椒醬	2 湯匙
糖	2 湯匙

做法

1. 用油爆香乾葱頭，盛起備用。
2. 倒入肉絲炒約 5 分鐘，在這過程中肉絲會出水，當水蒸發約 3 分鐘後，待汁收乾變成油，再把乾葱頭回鑊，加入調味料兜勻，煮 5 分鐘即可。

GIGI NOTES

炸醬隔一夜才吃味道更佳，調味可按個人喜好口味調整。
炸醬拌飯或拌麵享用都非常美味，圖中用的是蛋麵。

紅燒醉肉

Braised Pork Belly in Brown Sauce

材料

五花腩肉	2 大條（約 1 斤）（切大塊，汆水）
紅指天椒	1 隻
冰糖	1 湯匙（舂碎）
老抽	1 湯匙
蠔油	4 湯匙
紹興酒	1 杯半
水	半杯
鎮江香醋	1 湯匙（後下）

做法

1. 用約 2 湯匙油爆炒指天椒及冰糖，糖溶化後加入五花腩肉，兜炒至上色。
2. 倒入老抽、蠔油、紹酒和水，燜 60 分鐘，待汁液濃稠後加醋即可。

GIGI NOTES

經汆水後的豬肉減少了肉臊味，也減少了肥脂。

肥瘦相間的五花腩肉是最宜炮製這菜的食材，肥肉入口即溶，沒有肥膩的感覺。

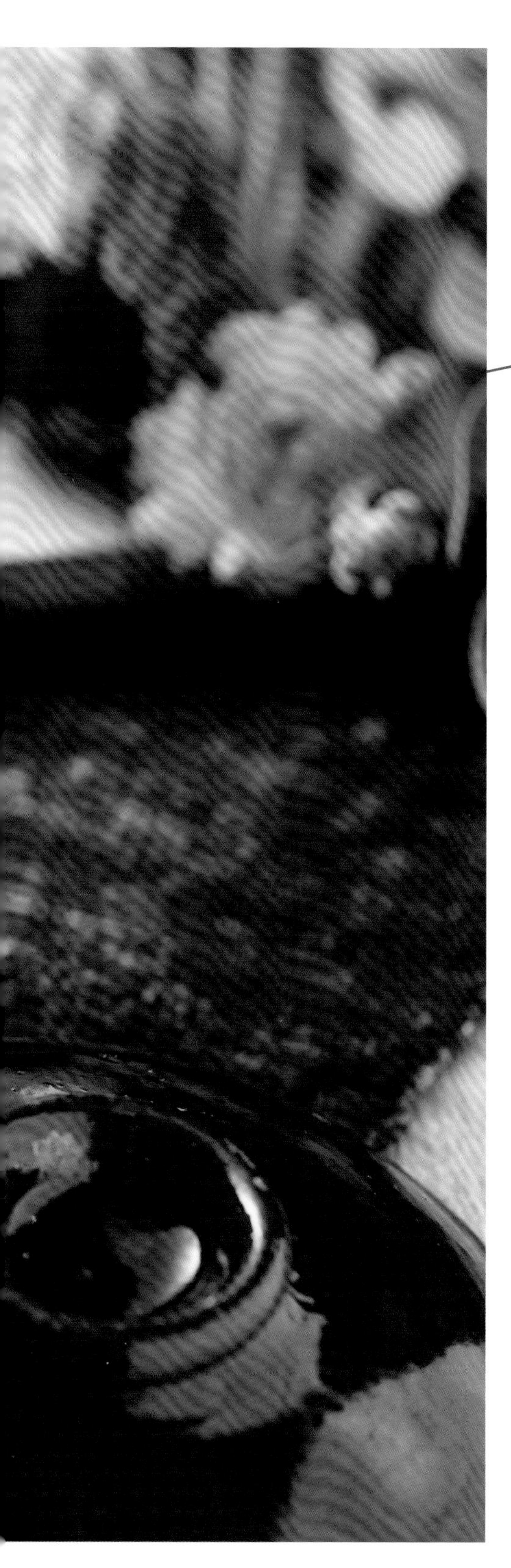

涼瓜黃豆鹹豬骨煲

Braised Salted Pork Ribs with Bitter Melon and Soybeans in Casserole

材料

排骨	半斤（用 1 茶匙鹽醃過夜）
苦瓜	2 個（去籽，切成角形）
黃豆	1 杯（浸過夜）
薑	2 片

做法

1. 洗去鹹排骨的鹽分。
2. 燒熱砂鍋，用油爆香薑片，加入排骨和黃豆，再加入滾水 2 杯，煮半小時至排骨腍。
3. 待剩餘少許汁液時，加入苦瓜煮 2 分鐘即可；如喜歡腍身的口感，可煮 5 分鐘。

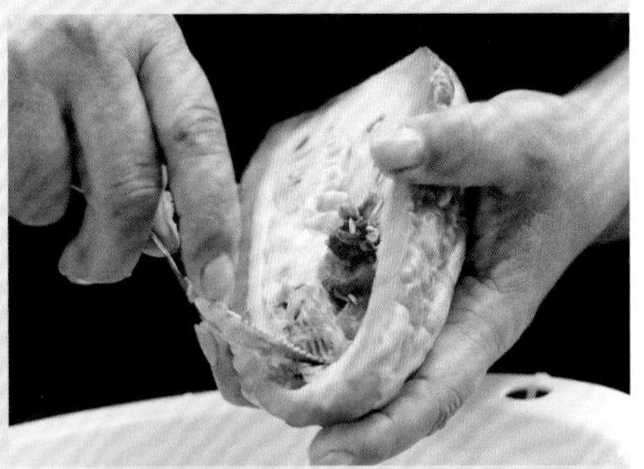

GIGI NOTES

經過鹽醃一夜的排骨，肉質結實，味道甘香，宜用來燜煮。

經過浸泡一夜的黃豆，質地變腍了，容易煮軟。

下滾水的原因是，令溫度不會驟然下降，維持燜煮時的溫度。

Allrounder 砧板以再生纖維製成紙，層疊加壓一起，並加入食品安全級樹脂生產而成，這種物料輕巧、穩定性高，表面無孔及耐污，亦不含金屬、揮發性或半揮發性的有機物質，由刀具造成的刮痕亦較淺，讓細菌不易藏在砧板中。

獅子頭

Braised "Lion Head" Pork Balls

材料

免治豬肉	1斤
黃芽白	1棵（約2斤）（切粗條）
薑	2片

醃肉料

鹽	1茶匙
酒	1茶匙
麻油	2茶匙
粟粉	2茶匙
胡椒粉	適量
葱花	2茶匙
薑米	1茶匙

封肉料

老抽	1湯匙	調勻
水	1湯匙	
粟粉	1湯匙	

調味料

鹽	1茶匙
糖	1茶匙
老抽	1湯匙

做法

1. 免治豬肉加入醃肉料，拌後撻打幾下，做成 4 個大肉丸，沾上封肉料，在砂鍋內煎至兩面微黃，小心取出肉丸。
2. 用原鍋內的油，爆炒薑片及黃芽白條，倒入 1 杯滾水，煮滾後加入肉丸，用慢火燜煮，加入調味料煮半小時即可。

GIGI NOTES

封肉料的作用是彷似一層保護罩，讓肉丸外層較硬，肉汁不會溢出，保持內裏濕潤。

黃芽白索滿豬肉的肉香，肉丸則很 juicy，這是一個超讚的伴飯餸。

話梅豬手

Braised Pork Trotter Plum-Scented Wine Sauce

材料

豬手	1 隻（斬大件，汆水）
大粒花生	4 兩

配料

薑	4 片
蒜頭	4 粒（拍碎）
乾葱頭	4 粒（切碎）
紅指天椒	1 隻
冰糖	1 湯匙（舂碎）

調味料

梅子	4 粒
大話梅	6 粒
紹酒	2 湯匙
老抽	1 湯匙

做法

1. 先爆香配料，加入豬手和花生，再下調味料和過面水。
2. 先用大火煮滾，再轉用中小火煮約 1 小時 30 分鐘，熄火後焗半小時即可。

GIGI NOTES

附圖中的配料如蒜頭、紅辣椒、乾葱頭等，是菜式重要的靈魂，有增香的作用。

它似蜜

Fried Beef Tenderloin in Sweet Sauce

材料

牛柳　300 克（切片）（醃半小時）
麻油　4 湯匙
青瓜仔　2 條（刨片，用半茶匙鹽醃 2 小時，榨乾水分，圍碟邊。）

牛柳醃料

水　2 湯匙
雞蛋白　2 湯匙
魚露　1 湯匙
生粉　2 茶匙

醬汁料

甜麵醬　2 湯匙
糖　1 湯匙
油　1 湯匙

芡汁料

鎮江醋　1 湯匙
老抽　1 湯匙
紹酒　1 湯匙
薑汁　2 茶匙
生粉　1 茶匙

做法

1. 用中火燒紅麻油，把牛柳煎至五成熟，取出備用。
2. 用中火把醬汁料煮至糖溶，牛肉回鑊，加入芡汁料，兜勻上碟，伴以青瓜片享用。

GIGI NOTES

傳統上它似蜜是採用羊肉，因為羊肉有臊味，故我用牛肉代替。

牛柳上可能會有筋，宜切去。

牛肉用蛋白醃，可保持肉質嫩滑。

蓮藕魚頭煲

Fish Head Casserole with Lotus Roots

材料

大魚魚頭	1 個（連少許魚身肉）（分切成 4 大塊，抹鹽和胡椒粉調味。）
薑	4 片

配料

蓮藕	1 小節（去皮、切薄片）
洋葱	半個（切塊）
芹菜	1 棵（切度）
蒜頭	8 粒（拍扁）

醬料

豆瓣醬	2 湯匙	調勻
甜麵醬	2 湯匙	
酒	2 湯匙	
薑末	1 湯匙	
葱末	1 湯匙	
鹽	半茶匙	
糖	1 茶匙	

GIGI NOTES

爆香薑片後才煎香魚頭，有辟去魚腥的效用。

吃魚頭較花時間，砂鍋有保溫的作用，就算吃得久一點，菜式也不會太快涼掉。

做法

1. 燒熱砂鍋下油爆香薑片，煎香魚頭，盛起備用。
2. 用餘油爆香配料，把魚頭放在上面，醬料倒在魚頭上，加入熱水半杯，煮 10 分鐘即可。

蠔烙 Fried Oyster Fritter

材料

蠔仔	300 克（加生粉洗淨，索乾，切粒。若用桶蠔要切小粒。）
乾葱頭	2 粒（切碎）（爆香備用）
芫茜	1 棵（切碎）
鴨蛋	2 個（拂勻）
幼粒番薯粉	3 湯匙
粟粉	1 茶匙
胡椒粉	適量
魚露	1 茶匙
油	半茶匙

做法

1. 把全部材料（除蠔粒、乾葱茸外）混合後拌勻。
2. 再加入蠔粒、乾葱茸混合，用慢火熱油煎香兩面即可。

GIGI NOTES

蠔仔多黏上碎殼和裙邊有點髒，宜用生粉揉去髒物和沖走碎殼，吃時才安心。

把蠔仔切粒的原因是令蠔較易煎熟；也可將蠔仔氽水至半熟才切粒。

宜用幼番薯粉，較容易拌勻成粉漿；番薯粉的作用是讓蠔烙微脆中帶點煙韌。

鴨蛋蛋黃的色澤較紅，蛋味較香，製作蠔烙多用鴨蛋。

白鱔上沙灘

Crispy Fried Eel Strips Covered in Cornflakes

材料

蒲燒鰻魚	1 條（切粗條）
粗粒番薯粉（地瓜粉）	適量
粟米片	1 杯（壓碎）

醬料

蛋黃醬	4 湯匙
煉奶	1 湯匙
山葵醬（日本芥末）	1 湯匙

* 全部混合

做法

1. 鰻魚條沾上番薯粉，炸至金黃色。
2. 取出沾上醬料，再沾上粟米片碎即食。

GIGI NOTES

粟米片可用米通代替。

番薯粉分為粗粒和幼粒兩種，粗粒用作裹料，沾滿食材後油炸，香口酥脆，脆度維持較久，因面上有脆粒的關係；而幼粒番薯粉宜做蠔餅。

這個小吃宜配上冰凍啤酒享用。

乾咖喱蟹 Dry-Fried Curry Crabs

材料

蟹	2隻（用牙刷擦淨，斬件撲上適量乾粉，走油。）
洋葱	半個
乾葱頭	2粒
蒜頭	2粒
牛油	1湯匙
生油	1湯匙
咖喱醬或咖喱膏 *	2湯匙
青、紅西椒	各半個（切碎，後下）
酒	少許

* 咖喱醬材料、做法見後頁

調味料

生抽	1湯匙
椰汁	1湯匙
花奶	1罐（小）

做法

1. 用牛油、生油爆香洋葱、乾葱頭和蒜頭後，下咖喱醬或咖喱膏2湯匙。
2. 將蟹回鑊，灒酒，與（1）兜勻後調味，最後把青紅西椒加入，兜勻即可。

咖喱醬

材料

洋葱	2 個（切碎）
蒜頭	4 粒（拍碎）
薑	4 片（拍扁）
紅辣椒	2 隻（剁碎）
芫茜	2 棵
黃薑粉	1 茶匙
肉桂粉	1 茶匙

配料

椰奶	1 罐（大）
水	半杯
鹽	1 茶匙
黑椒粉	1 茶匙

做法

全部材料放入攪拌機打成蓉，熱鑊下油將材料倒入鑊中用小火炒香，再加配料煮滾後即可。

GIGI NOTES

咖喱醬可照上述方法自製或購買現成的，以自己方便為原則。

桂魚片粉皮

Mung Bean Noodles with Mandarin Fish

材料

桂花魚	1 條（起肉、切片，醃半天）
粉皮	2 塊（浸至軟，撕碎）
上湯	1 杯

醃魚肉料

酒	2 茶匙	
薑	2 片	拍扁
葱白	2 條	拍扁
胡椒粉	適量	
生粉	1 湯匙	

配料

青、紅西椒	各半個（切塊）
薑	2 片

調味料

磨豉醬	3 湯匙	調勻
甜麵醬	2 湯匙	調勻
酒	2 湯匙	調勻
老抽	1 茶匙	調勻
糖	1 茶匙	調勻

做法

1. 熱鑊下油，用慢火爆香薑片，加入青、紅西椒，倒下調味料略煮一會，撈起配料。
2. 倒入上湯，待上湯煮滾後，放入魚片及粉皮。
3. 待魚片熟後，倒入配料即可享用。

GIGI NOTES

圖中用的是乾粉皮，要浸軟才用；你也可買新鮮的粉皮，可即用。

除了桂花魚外，可以用鯇魚脊起雙飛代替。

川味鯇魚腩

Seared Grass Carp Fillet in Sichuanese Sauce

材料

鯇魚腩	1塊（用鹽、胡椒粉略醃）

配料

薑	4片
蒜頭	4粒
大豆芽	4兩（去根）
半生熟木瓜	1個（去皮去籽、切塊）
海帶	1兩（浸軟、切度）
豆腐	1磚（切塊）

調味料

豆瓣醬	1湯匙
糖	1湯匙
生抽	1湯匙
滾水	2杯（分2次加入）

後下料

青蒜	1棵（切度）
紅辣椒	1隻（切片）
芫茜	1棵（切碎）

做法

1. 用熱鑊下油，待油燒熱後把魚煎至兩面金黃色。
2. 另鑊爆香薑、蒜頭，把配料順序倒入，同時加入調味料和一杯滾水，冚蓋略煮。
3. 把魚倒入，再加入餘下一杯滾水，滾後加入後下料，即可上桌。

GIGI NOTES

煎魚時要用熱鑊熱油，讓魚能迅速煎香定型，不致支離破碎。

後下料是取其香和顏色，不需要久煮。

焗中式火雞

Roast Turkey in Chinese Style

材料

火雞	1 隻（約 8 磅）（醃過夜）
欖子	2-3 個（搓爛）

配料

洋葱	1 小個	切碎
紅蘿蔔	半條	切碎
西芹	2 枝	切碎
香葉	2 片	

醃料（視乎火雞的大小）

鹽	2 茶匙
沙薑粉	2 茶匙
老抽	2 湯匙
紹興酒	2 湯匙

* 註：因製作時未能購買火雞，故暫用雞代替，但做法不變。

做法

1. 把櫻子塞入已醃味的火雞肚內。
2. 預熱焗爐 230℃。
3. 準備一個大焗盤，盤內放一架子，下放配料，倒入滾水 1 杯，上放火雞（雞胸向上）。
4. 先用猛火 230℃焗 10-15 分鐘，見胸部呈微黃，在雞身墊錫紙防燒焦，改用 170℃繼續焗，開始計時 8 磅火雞焗 96 分鐘（烤焗時間視乎火雞大小而定，每磅火雞以 12 分鐘計算，請自行計算所需時間。）
5. 熄火後，不要取出火雞，待在焗爐內吸餘溫半小時即可。

GIGI NOTES

火雞改用 170℃繼續烤焗的時間，並不計算用猛火焗的 10-15 分鐘，以及往後待在焗爐吸收餘溫之時間。

如用作上色，我會選用草菇老抽。

醃料的使用量，我會視乎火雞的大小而增添、減少。

將火雞放在三皇（洋葱、紅蘿蔔和西芹）和香葉，除可讓雞吸收香氣外，也可防止雞皮黏着焗盤。

香酥雞 Crispy Fried Chicken

材料

雞上髀肉	3 隻（切塊，醃過夜）
九層塔	4 棵（摘下葉，炸香拌碟用。）
粗粒番薯粉（地瓜粉）	半杯
淮鹽	適量（伴食用）

醃料

雞蛋	1 個
五香粉	1 茶匙
糖	2 茶匙
蒜泥	2 茶匙
紹酒	1 湯匙
生抽	1 湯匙

做法

1. 將醃過夜的雞肉灑上番薯粉，放入 180℃ 油中炸熟，取出備用。
2. 將油再加熱，把雞倒入用中火再炸一次，撈起即可，以淮鹽伴食。

GIGI NOTES

九層塔葉要抹乾才炸，如有水分，炸時會濺油。
雞肉蘸上番薯粉後宜即炸，否則番薯粉被浸濕後，炸後不夠酥脆。
宜用粗粒番薯粉，皮料才酥脆。
這道香酥雞有台灣夜市炸雞的風味。

椰菜燉雞

Braised Chicken with Cabbage

材料

雞　1 隻
乾螺頭　1 兩（處理過，做法看 P.134）
椰菜　1 個（一開三）
水　4 杯
米酒　1 杯
生抽　半杯
冰糖　2 粒
乾葱頭　4 粒
蒜頭　4 粒
京葱　1 棵（切度）

做法

1. 將 1 份椰菜墊鍋底內，雞放在上面。
2. 其餘 2 份椰菜放在雞旁邊，下其餘材料，燉 3 小時即可。

GIGI NOTES

若有上海年糕或韓國年糕，可汆水，加入煮至軟即可。

九層塔鮑魚雞

Stir-Fried Chicken and Abalone with Thai Basil

材料

雞上髀肉	4 大塊（切塊，調味）
罐頭鮑魚	1 罐（原罐燉 2 小時至腍，切塊）
冰糖	2 粒
洋葱	1 個（切塊）
薑	4 片
乾葱頭	4 粒（略拍）
紅辣椒	1 隻
九層塔	4 棵（只要葉）
麻油	2 茶匙
老陳醋	2 湯匙
魚露	1 湯匙

醃雞料

鹽	半茶匙	糖	1 茶匙
米酒	1 湯匙	黑椒碎	半茶匙

做法

1. 先用 2 湯匙油爆香雞肉至八成熟，盛起。
2. 再用 2 湯匙麻油爆香洋葱、乾葱頭、薑片和冰糖至有香味，加入雞肉及鮑魚，再下老陳醋、魚露兜炒 2 分鐘，加入九層塔兜勻即可。

GIGI NOTES

宜臨上碟時才下九層塔，才能保持香氣。

可試試原罐燉鮑魚：用過面水煮原罐鮑魚 2 小時，煮後撈出來攤涼才開罐。這時鮑魚的質感腍滑可口。

因雞及鮑魚的烹調時間有別，故兩者需預先個別處理，之後再混合一起烹煮。

麻辣臘味燒蘿蔔

Braised White Radish with Preserved Pork Belly in Mala Sauce

材料

臘肉	半條（汆水後斜刀切片）
蘿蔔	1 條（去皮、切粗條）
京葱	1 棵（斜刀切片）
蒜頭	8 粒
紅椒	1 隻（斜刀切片）
花椒粒	1 湯匙（冷鑊下油，放入花椒粒和一半紅椒片，炸至花椒粒和紅椒變色出味後，撈起花椒粒和紅椒，即成花椒油。）

調味料

蠔油	1 湯匙	調勻
老抽	1 茶匙	
片糖	1/8 塊（舂碎）	
胡椒粉	少許	
滾水	1 杯	

做法

1. 用花椒油爆炒蒜頭、餘下紅椒片，至香氣四溢後，加入蘿蔔和臘肉兜炒。
2. 倒入調味料，用大火收乾水分後，加入京葱，兜勻即可享用。

GIGI NOTES

臘肉需要汆水的原因是，除了可去油脂外，也較容易切成片。

將油爆後的花椒棄去，是避免當你大快朵頤時，突然吃到硬硬的顆粒。

宜用冷鑊下油爆炒花椒，花椒會逐漸出味；如將鑊燒熱，油大滾時下花椒，花椒會易燶和變苦。

紅燜蘿蔔

Braised White Radish

材料

白蘿蔔	2 條（去皮，切滾刀塊）
上湯	2 杯
薑	2 片
片糖	1/4 塊（舂碎）

做法

1. 熱鑊下油，爆炒薑片、片糖，放入白蘿蔔兜勻，倒入上湯，燜約 1 小時。
2. 取出，放在碟上，以餘下之汁料打芡，淋上蘿蔔即可。

GIGI NOTES

白蘿蔔宜切成滾刀塊，此切法適合做燉煮的烹調方法。

宜購買「墜手」的白蘿蔔，代表水分足、肉質嫩。可用削皮刨刨去薄薄的外皮，不會因用刀不當削去過多的蘿蔔肉造成浪費。

可以把全部材料放入壓力鍋煮半小時，節省不少時間。

糖的作用除了可調味外，也可以令食材軟化，所以牛腩、牛脹等都會放糖燜煮。

出版這書適逢夏季，炮製白蘿蔔菜式時，多買一條白蘿蔔試試炮製開胃的「酸甜蘿蔔皮」吧。準備白蘿蔔、添丁甜醋、白醋、山西陳醋、鮮醬油、砂糖和指天椒。白蘿蔔不去皮，切成厚片，用 1 湯匙鹽醃半小時後，即可榨乾水分。蘿蔔連指天椒放入醋料內，泡浸半日即可。

Victorinox 的 Swiss Modern 廚刀系列，以極致耐磨的不銹鋼製成，精準鋒利的刀刃令切肉剁菜變得輕而易舉，並令切面平滑，更可靈活處理各樣細微的工序，入廚得心應手。

豬皮娃娃菜

Braised Baby Napa Cabbages with Pork Skin

材料

娃娃菜	3 棵（一開二，切粗條）
冬菇	8 朵（已處理，做法看 P.135）（切粗條）
薑	4 片（拍爛）
乾豬皮	1 大塊（浸軟後切塊，用薑葱加醋汆水，擠乾水分）

調味料

蠔油	1 湯匙
老抽	1 茶匙
胡椒粉	1 茶匙
麻油	1 茶匙

做法

1. 熱鑊下油，用薑片爆炒娃娃菜、豬皮和冬菇，加入熱水 1 杯。
2. 用中大火煮 15 分鐘，加入調味料兜勻即可。

GIGI NOTES

市場有已浸發的急凍豬皮售賣，汆水後即可使用。

乾豬皮在雜貨舖有售。乾豬皮買回來後，用凍水浸軟後切塊，汆水；汆水的目的是去除豬皮的油膩和臊味。

醋溜椰菜
Stir-Fried Cabbage in Spicy Vinegar Sauce

材料

椰菜	1 小個（撕成大塊）
花椒粒	1 湯匙
乾小紅椒	4 隻（剪碎）
薑米	2 湯匙

調味料

香醋	2 湯匙
糖	2 湯匙
生抽	1 湯匙
鹽	半茶匙
生粉水	適量（半杯水加生粉 2 湯匙調勻）

做法

1. 用 2 湯匙油慢火爆香花椒粒及一半小紅椒碎，出味後撈走。
2. 把剩餘的小紅椒碎及薑米爆炒後，轉大火加入椰菜，再加入調味料，不停兜炒至僅熟、仍脆口即上碟。

GIGI NOTES

可以用黃芽白代替椰菜，宜兜炒至軟身才上碟；而椰菜則食爽口的。

如用黃芽白，宜切成粗條才炒，以便受熱更快。

蓮藕豆豉辣椒粒

Stir-Fried Diced Chillies with Lotus Roots and Fermented Black Beans

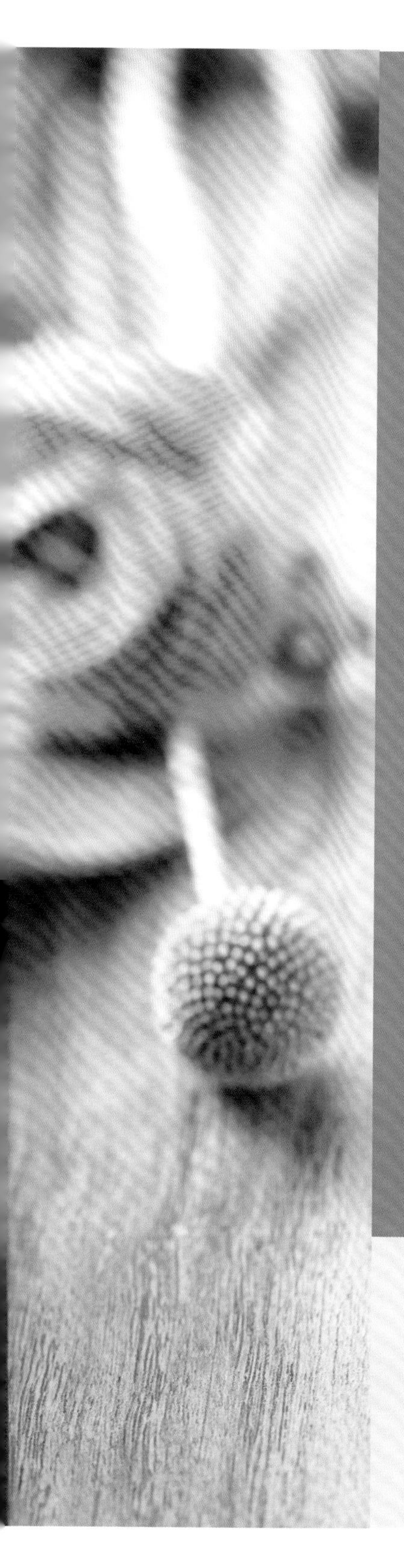

材料

蓮藕	1 節（切丁，泡水去澱粉，沖洗幾次，汆水時下少許糖。）
長青辣椒	2 隻（切粒）
長紅辣椒	1 隻（切粒）

配料

薑	4 片（切幼粒）
蒜頭	4 粒（切成蓉）
豆豉	2 湯匙（略剁，加糖和油混合。）

調味料

生抽	2 湯匙
糖	1 湯匙
醋	1 湯匙

做法

1. 爆炒蓮藕粒，加入配料，兜炒至入味。
2. 加入青紅辣椒粒，炒一會後加入調味料兜勻即可。

GIGI NOTES

可按個人喜好加入肉末。

如蓮藕不洗去澱粉，炒時會黏成一團；如是燜餸，則宜有澱粉。

蓮藕汆水時下少許糖，可辟去菜青味。

炒雙花

Stir-Fried Broccoli and Cauliflower

材料

椰菜花　　半個（切小朵）
西蘭花　　1 棵（切小朵）
蒜頭　　3 粒（剁茸）
乾紅椒　　1 隻（切碎）
鹽　　1 茶匙
糖　　1 茶匙
麻油　　適量

做法

1. 椰菜花、西蘭花用 1 湯匙油和 1 茶匙鹽汆水 2 分鐘，瀝乾水分。
2. 用中火爆炒蒜茸及紅椒，出味後（但不要焦）轉大火，加入已汆水的雙花，下少許鹽、糖及麻油，兜勻即可上碟。

GIGI NOTES

\# 將已切成小朵的西蘭花，放入下了 1 至 2 茶匙鹽的水內，浸約 15 分鐘，隱藏在花內的小蟲便會浮出水面，跟着再沖洗一次，洗好後便可以煮了。

\# 椰菜花、西蘭花的質地較硬，在炒前先汆水，讓它較腍身，可縮短兜炒的時間。

黃金粟米

Sweet Corn Kernels in Salted Egg Yolk Sauce

材料

新鮮粟米粒	3 條份量
鹹蛋黃	8 個（蒸熟，趁熱搗碎）
紅西椒	半個（切粒）
青西椒	半個（切粒）
牛油	1 茶匙
油	1 茶匙
紹酒	1 湯匙
鹽	半茶匙
糖	半茶匙
黑椒碎	適量

做法

1. 用慢火加熱油，下牛油，待牛油溶化後，放入鹹蛋黃碎炒至起泡，灒酒。
2. 放入粟米粒後轉大火，下鹽和糖兜炒約 2 分鐘，加入青、紅西椒粒略兜炒，下黑椒碎兜勻即可上碟。

GIGI NOTES

一定要用新鮮粟米粒，勿用罐頭或冷藏貨色，除了會出水、口感差之外，更沒有新鮮粟米粒咬下去爆出汁液的甜美感覺。

牛油和鹹蛋黃是絕配，只有用牛油炒鹹蛋黃才能帶出它的香。因牛油易炒燶，故要在鑊內先下少許油，才放牛油，就可以避免炒焦的情況出現。

剩下來的粟米芯可放在雪櫃冷藏，日後用來煲湯，不要浪費。

豉椒炒萵筍

Stir-Fried Celtuce in Chilli Black Bean Sauce

材料

萵筍	1 棵（去皮、去筋，切粗條，用 1 茶匙鹽略醃。）
豆豉	1 湯匙（熱水略浸）
蒜茸	1 湯匙
薑米	1 湯匙
紅辣椒	1 隻
紹酒	1 湯匙
糖	1 茶匙
麻油	適量

做法

1. 燒熱鑊下油，爆香薑米、蒜茸、豆豉，灒酒。
2. 加入萵筍、紅椒，快手兜炒，下糖和麻油炒勻即可。

GIGI NOTES

用 1 茶匙鹽略醃萵筍的作用，是讓萵筍出水和口感爽脆。

用熱水略浸豆豉的作用，是讓豆豉質感軟些，炒時容易出味。

雞肉鮮果沙律

材料

雞胸肉	2 塊（斜刀切薄片，醃 2 小時）
青瓜	1 條（刨長條）
菠蘿	半個（切條）
草莓	8 顆（一開四）
藍莓	1 盒

醃雞料

鹽	1 撮
檸檬汁	半茶匙
生粉	1 茶匙
橄欖油	1 茶匙

沙律汁

鹽	半茶匙	調勻
糖	1 茶匙	
檸檬汁	1 茶匙	
薄荷葉	10 片（舂碎）	

做法

1. 煲水至八成熱，見鍋底有泡泡即放鹽 1 茶匙、油 1 湯匙，把雞肉逐少放入，每片灼約 10 秒鐘後取出。
2. 把所有材料與沙律汁拌勻，即可享用。

GIGI NOTES

雞肉灼至僅熟即可，才能享受嫩滑的口感。

Fruit Salad with Chicken Breast

營養早餐麥餅

Healthful Breakfast Oatmeal Cookies

材料

室溫牛油	1/4 磅（或用微波爐叮 30 秒）	
紅糖	半杯	
大雞蛋	1 個（拌散）	
麵粉	3/4 杯	篩勻
發粉	半茶匙	篩勻
肉桂粉	半茶匙	篩勻
鹽	1 撮	篩勻
即食麥皮	1 杯半	
提子乾	半杯	

做法

1. 將牛油、紅糖和雞蛋拌勻後，加入已篩勻的粉類，拌勻，加入麥皮和提子乾，拌勻成粉糰。
2. 將粉糰分做成小圓餅，放在已預熱 170℃（350 ℉）焗爐焗 15 分鐘，反轉餅身再焗 10 分鐘即可。

GIGI NOTES

可以加入合桃碎，味道香兼更有營養。

焗後攤涼後的營養早餐麥餅，宜放入密封的保鮮盒保存。

我以這麥餅為早餐，每次我都會烘焙一星期的份量，可伴牛奶或掰碎放入乳酪內都好好味。

這個麥餅的做法簡易，不需要特別的技巧就能成功。

臘腸鬆餅

Baked Cantonese Pork Sausage Pancakes

材料 A

凍牛油	1/8 磅（切細細粒）	
麵粉	1 杯	篩勻
糖	1 湯匙	
鹽	1/4 茶匙	
梳打粉	1/4 茶匙	
蒜粉	半茶匙	
發粉	1 茶匙	

材料 B

牛奶	1 杯（8 分滿）
雞蛋	1 個（打散，連同牛奶共 1 杯）

材料 C

臘腸	1 條	切細粒
葱花	1 條	
芝士碎	1 杯	

做法

1. 預熱焗爐 180℃。
2. 將全部材料 A 混合，用打蛋器打至牛油溶化約 1 分鐘。
3. 加入材料 B 繼續攪拌均勻，最後倒入材料 C，改用膠刮刀混合成糊狀。
4. 焗盤上鋪上烘焙紙，舀上適量麵糊，放入已預熱焗爐焗 20-25 分鐘即可。

GIGI NOTES

這是一個結合中西方食材與做法的鬆餅：臘腸的濃香配上葱花和芝士碎，出乎意料地相配，配上熱茶，是下午茶的首選。

炮製這鬆餅最關鍵的步驟是要將牛油切碎，用打蛋器打時才容易溶掉。

舀上適量麵糊在焗盤時，麵糊與麵糊之間要留些空間，因為麵糊在烘焙時會脹大。

椰香奶凍 Coconut Panna Cotta

材料

牛奶	3/4 杯
淡奶	1/3 杯
椰漿	2/3 杯
糖	1/4 杯
魚膠粉	2 茶匙（用 4 湯匙溫水調勻）

做法

1. 把以上材料（除魚膠粉水外）放入小鍋內，慢火煮至糖溶。
2. 再把魚膠粉水倒入，拌勻後熄火，冷卻後放入雪櫃內凝固即可。

GIGI NOTES

宜先用溫水調溶魚膠粉才倒入奶凍料內，如直接灑進魚膠粉，魚膠粉會結塊，影響凝固的效果。

奶凍料煮至微滾、糖溶就可以了，不需要大滾，否則奶凍料質感會粗糙。

桂花銀杏

Candied Gingkoes with Osmanthus

材料

銀杏	2 杯（去殼、去芯）
冰糖	半杯
桂花醬	1-2 湯匙

做法

1. 銀杏、冰糖加到過面水煮滾，收中火拌勻。
2. 煮至銀杏微脸、水分收乾，即可加入桂花醬，拌勻即成。

GIGI NOTES

桂花醬可在南貨舖購買。

銀杏不宜多吃，這甜點小吃宜與朋友分享。桂花銀杏口感煙韌、味道微甘帶甜。

紅豆南瓜餅

Pumpkin Pancakes with Red Bean Filling

材料

日本南瓜	半個（約600克）（去籽，蒸15分鐘起肉，趁熱用叉壓成茸。）
糯米粉	300 克
糖	3 湯匙
紅豆蓉	適量
芝麻	適量

做法

1. 把糯米粉倒入南瓜蓉中，連糖一齊搓成糰。
2. 取出適量麵糰，搓圓壓扁，包入適量紅豆蓉，收口，再壓扁後，面塗少許水，壓向芝麻，稍微用力使芝麻沾滿餅面。
3. 燒熱平底鑊，下油，用慢火煎至兩面金黃即可。

香蕉雪芳蛋糕

Banana Chiffon Cake

材料 A

雞蛋白	7 個（用中高速打約 3 分鐘至企身）

材料 B

香蕉	3 隻（用叉子壓爛）
牛奶	半杯
菜油	半杯
雞蛋黃	7 個
糖	半杯

材料 C

麵粉	1 杯半	篩勻
發粉	1 湯匙	
鹽	1 小撮	

做法

1. 預熱焗爐 160℃（305 ℉）。
2. 將全部材料 B 混合略打，隨即將材料 C 逐少加入打勻。
3. 將材料 A 倒入材料 B 內，輕手拌勻。
4. 放入焗爐以 160℃（305 ℉）焗 1 小時，取出倒扣至涼透，切件享用。

Francfranc

GIGI NOTES

一定要採用呈現梅花點的香蕉，蕉肉香甜又軟糯，適宜用來烘焙。

香蕉不要放入雪櫃內，否則蕉肉會變黑。

用保鮮紙包着香蕉柄，可以保鮮。

宜用附圖的烘焙模，將已烘焙好的蛋糕反轉散熱，否則蛋糕會隨地心吸力下陷。

雪芳蛋糕不需要下發粉等膨脹劑，而是靠打發蛋白內的空氣遇熱膨脹而發起。

果醬蛋糕

Fruit Jam Cake

材料 A

室溫無鹽牛油	225 克
糖	半杯
雞蛋（室溫）	4 個
果醬	4 湯匙

材料 B

麵粉	1 杯半
發粉	1.5 茶匙
鹽	1 小撮

做法

1. 預熱焗爐 170℃。
2. 材料 A：用打蛋器高速打牛油 1 分鐘，分兩次倒入糖，待糖溶化後將雞蛋逐個加入（待每個雞蛋充分打勻時，才加入另一個），最後拌入果醬。
3. 材料 B：把全部材料混合篩勻，分兩次倒入材料 A 的混合物內，慢速打勻成蛋糕漿料。
4. 將蛋糕漿料倒入焗盆內，放已預熱焗爐焗 40 分鐘即可。

GIGI NOTES

可以用任何果醬，圖示的是用草莓果醬，可用鮮草莓伴食；如果用藍莓果醬，亦可加入鮮藍莓同焗。

做法 2 中特別說明每個雞蛋與牛油糖混合物充分打勻融合後才加入另一個雞蛋，這樣可避免出現「豆腐渣」油蛋分離的情況，讓油分沉在糕底，令蛋糕「發」得不好。

意大利方塊麵

Classic Lasagne in Meat Sauce

材料

方塊麵	1 盒
免治牛肉	1 磅
免治豬肉	半磅
混合芝士碎	適量（後下）

調味料

黑胡椒碎	1 茶匙
鹽	1 茶匙

配料

洋葱	1 個（切碎）
番茄	3 個（切碎）
番茄肉	1 罐（400 克）
蘑菇	半磅（切碎）
香葉	2 片
片糖	半塊
番茄醬	2 湯匙
番茄膏	2 湯匙

做法

1. 白鑊烘炒免治牛肉和免治豬肉至水份收乾，加入調味料炒勻後盛起。
2. 另鑊用牛油和生油爆炒香洋葱至透明後，加入其餘配料，慢火煮 30 分鐘。
3. 加入免治牛肉和免治豬肉，略煮。
4. 在焗盤內先排一層方塊麵，舀入肉料，再排一層方塊麵，再舀入肉料，梅花間竹地鋪排，最後在肉上加入芝士碎。
5. 將焗盤放入已預熱 190℃（375℉）的焗爐內，焗 40-50 分鐘即可。

GIGI NOTES

在超市有售的生辣肉腸，剝去腸衣，內裏的免治肉可代替食譜內的豬肉和牛肉。因生辣肉腸內的肉已調味，故要試味才下鹽。

大蝦燜油麵

Braised Oil Noodles with Fried Prawns

材料

油麵	半斤
大蝦	半斤（大蝦用鹽抓洗 2 次，索乾水分，加胡椒粉醃片刻。）
蝦湯	1 杯
四季豆	4 兩（去老筋，斜刀切片，走油，撈起。）
雞蛋	2 個（拂勻，煎香，切成蛋絲。）
薑	2 片（切絲）
蒜頭	2 粒
鹽	半茶匙

做法

1. 大蝦起雙飛，去腸洗淨，瀝乾水分，泡油後撈起備用。
2. 爆香薑絲、蒜頭，倒入蝦湯煮滾，放入油麵，煮 2 分鐘，加入四季豆、蛋絲和鹽，再煮 1 分鐘。
3. 加入大蝦兜勻即可。

GIGI NOTES

每次我都將蝦頭、蝦殼收集起來，再煮成蝦湯，加了蝦湯的菜餚特別鮮美。

可將蝦頭、蝦殼用牛油爆香，加 1 杯水煮 3 分鐘，湯水呈橙紅色，隔渣備用，做成蝦湯。

菜式除了着重味道外，還要顧及賣相，宜先剖開蝦背，再挑去蝦腸。

魚頭雜豆湯

Fish Head Soup with Assorted Bean

材料

豬脹	半斤（汆水）	
大魚頭	1 大個（約 1 斤重）（加薑片煎香）	
腰果	半杯	全部用水浸半小時
花生	1/3 杯	
眉豆	1/3 杯	
黑豆	1/3 杯	
黃豆	1/3 杯	
陳皮	1 角	
滾水	15 杯	
紅棗	4 粒（去核）	

做法

1. 魚頭在鍋內煎香後，倒入滾水 15 杯，再把全部材料倒入。
2. 用大火煮滾 10 分鐘，改中火煲 3 小時即可。若用壓力煲，只需 10 杯水煮半小時即可飲用。

GIGI NOTES

\# 用薑片煎香魚頭的作用是去腥。

\# 倒入滾水的用處，是不會令爐溫驟然下降，增加烹調時間。

粉葛牛蒡墨魚鯪魚湯

Dace Soup with Kudzu, Burdock and Cuttlefish

材料

粉葛	2 斤（去皮，切成骨牌般大小）	
牛蒡	1 斤（略刮皮，切厚片）	
乾土茯苓	1 兩	
赤小豆	1 兩	
扁豆	1 兩	浸約 1 小時
穀芽	1 兩	
生麥芽	1 兩	
鯪魚	1 條（加薑片煎香）	
墨魚乾	1 隻（去墨魚骨後浸軟，切大塊）	
薑	5 片	
蜜棗	3 粒	
滾水	18 杯	

（乾土茯苓、赤小豆、扁豆、穀芽、生麥芽：浸約 1 小時）

做法

1. 將魚加薑片煎香，注入滾水，再放入其餘材料。
2. 先用大火煲滾 10 分鐘，再用中火煲 3 小時即可飲用。

GIGI NOTES

以紋理清楚的「老身」粉葛煲湯為佳，粉葛味香濃。宜將粉葛切成骨牌般大小，易出味。

如用壓力鍋，水可略減，只需 13 杯水煲約 1 小時即可。

這湯有健脾瀉火、消食和胃、清除口氣的功效。

飲用這湯時要小心魚骨；你可將魚放在煲湯魚袋，可免除鯁骨的危險。

七彩酸辣羹

Seven-Coloured Spicy and Sour Thick Soup

材料

水	5 杯
白蘿蔔	半條（刨絲）
蒟蒻絲（芋絲）	1 盒（剪短）
木耳絲	1 湯匙（浸軟）
冬菇	4 朵（浸軟，切絲）
紅蘿蔔	半條（刨絲）
柳梅	60 克（切絲） （用鹽、胡椒粉略醃）
中國芹菜	1 棵（切度，後下）

調味料

鹽	1 茶匙
糖	1 茶匙
香醋	1 湯匙
老抽	1 湯匙
胡椒粉	適量

生粉芡（拌勻）

生粉	2 湯匙
水	半杯

做法

1. 水煮滾後，除豬肉絲及芹菜段外，倒入其餘材料，滾約 5 分鐘，撇去浮泡。
2. 放入肉絲，煮約 1 分鐘後下調味料，滾後勾芡，灑上芹菜段即成。

GIGI NOTES

因為豬肉已切成絲容易熟，後下可讓豬肉保持鮮嫩；同一道理，最後才下芹菜，令芹菜保持翠綠和芹香。

楊桃蘋果雞湯

Chicken Soup with Starfruits and Apples

材料

雞	1 隻（汆水後去皮）
蛇果	4 個（一開四、去核）
楊桃	4 個（一開三、去核）
南北杏	2 湯匙
水	18 杯

做法

把全部材料放入鍋中，先用大火煲 10 分鐘，再用中火煲 3 小時即可。

GIGI NOTES

煲蘋果湯宜採用蛇果，因它清甜多汁，沒有其他品種的酸味。

將雞去皮才煲湯，讓湯有雞香而毫不肥膩。

沙參玉竹蜜瓜湯

材料

蜜瓜	1個（約2磅）（洗淨、去籽，連皮切大塊。）
瘦肉	半斤（汆水）
雞腳	8隻（汆水）
乾螺頭	4兩（已處理，做法看 P.134）
瑤柱	3粒
沙參	1兩（浸洗）
玉竹	1兩（浸洗）
無花果	4個（略洗，在中央剪一刀）
水	15杯

做法

把全部材料放入鍋中，先用大火煲10分鐘，再用中火煲3小時即可。

GIGI NOTES

在無花果中央剪一刀，會容易出味。

用雞腳煲湯有畫龍點睛之效，湯味香濃，入口充滿膠質。

Honeydew Melon Soup with Sha Shen & Yu Zhu

花旗參石斛瘦肉湯

Lean Pork Soup with American Ginseng and Shi Hu

材料

瘦肉	1 斤（汆水）
花旗參	1 兩（略浸）
石斛	1 兩
百合	1 兩（浸洗）
海竹頭	1 兩
無花果	1 兩（略洗，在中央剪一刀）
乾螺頭片	3 兩（已處理，做法看 P.134）
水	18 杯

做法

把全部材料放入鍋中，先用大火煲 10 分鐘，再用中火煲 3 小時即可。

GIGI NOTES

這湯有潤肺的功效，宜經常捱夜、煙酒過多的人士飲用。

花旗參能補氣、養心、提神；百合具有養陰潤肺、清心安神之功效；石斛則能滋陰、養顏、生津。

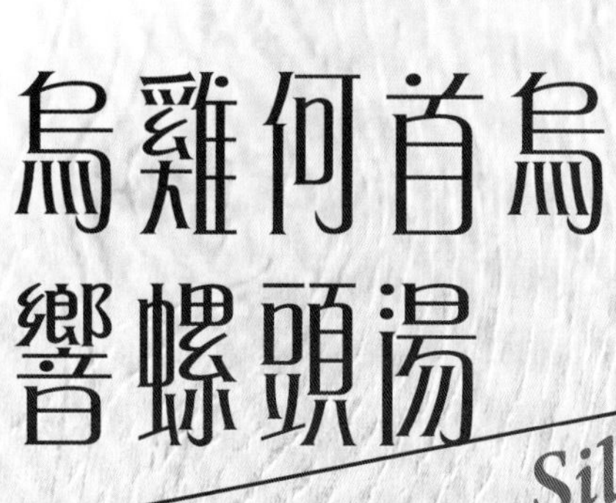

烏雞何首烏響螺頭湯

Silkie Chicken Soup with He Shou Wu and Conches

材料

烏雞	1 隻（汆水）	
何首烏	1 兩	略沖洗
當歸頭	1 兩	略沖洗
土茯苓	1 兩	略沖洗
黑豆	1 兩	浸約 2 小時
大粒花生	1 兩	浸約 2 小時
紅棗	1 兩（去核）	
乾螺頭	1 兩（已處理，做法看 P.134）	
陳皮	1 個（浸軟）	
薑	5 片	
水	18 碗	

做法

全部材料倒入鍋中，先用大火煲 10 分鐘，再用中火煲 3 小時。

GIGI NOTES

這湯用的湯料大多能補肝腎、養血，對預防脫髮及烏髮有功效。

雞保留雞皮煲湯，雖然湯會較油膩，但味道會較香。宜準備一個隔油湯壺，可將油脂隔除。

猴頭菇胡椒豬肚湯

Pork Tripe Soup with White Peppercorns and Monkey Head Mushrooms

材料

猴頭菇　1 兩（浸至發透）

白胡椒粒　1 兩（略壓碎）

白蓮子　3 兩（浸約 1 小時）

豬肚　1 個（洗淨去肥脂，飛水）

蠔豉　2 兩（浸透，洗去雜質）

薑　5 片

水　18 杯

做法

1. 全部材料倒入鍋中，先用大火煲 10 分鐘，再用中火煲 3 小時。
2. 如用高壓鍋煲湯，水可略減，用 13 杯水煲 1 小時即可飲用。

GIGI NOTES

街市豬肉檔有已清洗乾淨的豬肚售賣，但數量不多，宜早一天預訂。

蠔豉浸透後宜仔細清洗，因蠔身可能會黏附蠔殼，蠔裙也會暗藏污垢。

白胡椒粒有溫中下氣、消痰解毒之效；煲湯前將白胡椒粒略壓碎，會更易出味。

猴頭菇能健胃補虛、抗潰瘍、抗腫瘤。

這湯有開胃健脾、暖胃養胃、行氣消食的功效。

這食譜所用的 WOLL 高速微壓鑽石鍋（Low Pressure Cooker）內之氣壓始終比一般鍋具強，燜煮時間可縮減高達 50%，及節能高達 70%，既慳錢又環保。烹煮途中可隨時開蓋加入食材，毋須浪費時間等待排氣。

竹笙粟米羹

Cream of Sweet Corn Soup with Bamboo Fungus

材料

竹笙	6 條（浸軟、洗淨、切圈）
雞胸肉	1 塊（切小粒，用鹽、胡椒粉略醃）
葱絲	適量

湯料

雞湯	1 罐（約 250 毫升）
粟米蓉	1 盒（約 415 克）
水	3 杯

芡料

馬蹄粉	2 湯匙	調勻
水	半杯	

做法

1. 先將湯料煮滾，放入竹笙略滾；再把雞胸肉加入，一滾即可埋芡。
2. 撒上葱絲和胡椒粉即可。

GIGI NOTES

若要有嚼口，可加入 1 杯新鮮粟米粒。

蘋果忌廉湯

Apple Cream Soup

湯底材料

蘋果	5 個（一開四，去芯）
西芹	2 支（撕去老筋，切碎）
紅蘿蔔	1 條（去皮，切碎）
洋葱	3/4 個（切碎）
香葉	2 片
水	4 杯

湯料

蘋果	7 個（去皮、去芯，切粒）
馬鈴薯	1 個（去皮，切粒）
全脂奶	1 公升

伴食

忌廉	適量
肉桂粉	少量

做法

1. 先爆炒三皇（西芹、紅蘿蔔和洋葱）、香葉、蘋果，加入水，用慢火煮 1 小時，隔渣，湯底留用。
2. 把湯料倒入加熱去渣的湯底內，慢火煮 45 分鐘；離火後用手持攪拌器打至幼滑，澆下忌廉及灑少許肉桂粉在湯面即可。

GIGI NOTES

西餐中常用三皇：西芹、紅蘿蔔和洋葱來增加食物的香氣，例如 P.44 的焗中式火雞。

日月魚杞子明目湯

Lean Pork Soup with Moon Scallops & Goji Berries

材料

瘦肉	1 斤（汆水）
日月魚	1 兩（略浸）
杞子	1 兩 ⎤ 沖洗乾淨
淮山	1 兩 ⎦
粟米	2 條（一開四）
紅蘿蔔	2 條（去皮、切塊）
陳皮	1 個（浸軟）
蜜棗	3 粒
水	18 杯

做法

1. 水煲滾後，倒入全部材料。
2. 用大火煲 10 分鐘，改用中火煲 3 小時即成；若用壓力煲只需 13 杯水煲 1 小時即可。

GIGI NOTES

這湯有益肝、補肺、明目的功效。

丁香魚欖角辣椒醬

Chilli Sauce with Dried Herring and Salted Olives

材料

油	4 杯
蒜茸	2 杯
紅辣椒仔碎	2 杯
櫻花蝦	2 杯（略沖洗，瀝乾水，攪碎）
丁香魚	3 杯（略沖洗，瀝乾水）
欖角	1 杯（用熱水浸 5 分鐘撈起，剁碎）

調味料

米酒	4 湯匙
生抽	4 湯匙
糖	5 湯匙

做法

1. 熱鑊，用中火順材料次序各炒 3 分鐘。
2. 逐樣計時炒至丁香魚時，加調味料再炒 3 分鐘後加入欖角碎，繼續再炒 2 分鐘即可。
3. 冷卻後，放入玻璃瓶內儲存。

GIGI NOTES

盛醬料的玻璃瓶宜用熱水消毒，再瀝乾才使用。

玻璃瓶內的油一定要浸過材料面，可隔絕發霉菌。

嘉幹三文治醬

Willie's Garlic Sandwich Spread

材料

忌廉芝士	125 克
牛油	60 克
乾羅勒	1 湯匙
黑椒碎	1 湯匙
鹽	半茶匙
蒜茸	1 個蒜頭份量

做法

把全部材料拌勻，用大火蒸 10 分鐘，冷卻後即可食用。

GIGI NOTES

這三文治醬由舍弟嘉幹提供的。某天他興致勃勃地與我通電話，說發明了一個美味的抹醬，還提供了食譜，經我試做後，果然非常美味，宜用來塗烘厚麵包，伴牛油果、火腿食用。

建議先試味才下鹽。

乾螺頭的處理方法

1. 乾螺頭洗淨後，用過面水浸過夜，連水倒入電飯煲中，加 1 粒冰糖及 2 片薑，按掣煮至乾水，就可採用。
2. 放涼後，分別裝在塑膠午餐袋內，一份份置於冰格，需要用時，隨時取出，不必解凍即可烹調。

冬菇的處理方法

1. 將半斤冬菇沖洗乾淨，用水浸泡至軟身，去蒂，可留作煲湯用。
2. 冬菇和浸冬菇水加入薑2片、葱2條、片糖半塊，用保鮮紙封好，再刺兩個小孔，放入微波爐叮 10 分鐘。
3. 撈出冬菇，用2湯匙油爆香2粒蒜頭，下冬菇，加入2湯匙老抽，再倒入冬菇水，猛火將汁液煮至濃稠，加入蠔油2湯匙，關火。放涼後，分別裝在塑膠午餐袋內，一份份置於冰格，需要用時，隨時取出，不必解凍即可烹調。

Thank you!

謝謝您們，
讓這食譜的內容更豐富！

吳政栓中醫師

香港浸會大學中醫碩士；香港中文大學中醫進修文憑；
香港中文大學中醫骨傷科文憑。

食譜內大部分湯水，都是由吳政栓中醫師提供。

吳政楷

溫哥華加美參海味店老闆，中醫師吳政栓長兄。

他提供的竹笙粟米羹，簡易有益又美味。

老大梅基偉

他提供的炸醬，拌飯或拌麵享用都非常美味。

嘉幹

他是我的四弟，嘉幹三文治醬是他的新發明，宜用來塗烘厚麵包，伴牛油果、火腿食用。

陳國強

大廚陳國強師傅提供川味鮀魚腩，吃過的人無不讚好。

甄太

傳統二十四孝母親，四十八孝祖母。

為了孫兒，放棄自己退休安逸的日子，飛去三藩市陪同孫兒成長。她提供的營養早餐麥餅，是我的早餐必然之選。

Glutinous Rice Pork Balls with American Ginseng

INGREDIENTS

300 g half-fatty pork (sliced, rinsed, then finely chopped)
2 g American ginseng (sliced, soaked in water till soft, finely chopped; the soaking water saved for later use)
4 water chestnuts (peeled, put into a plastic bag, crushed with the flat side of a knife and then finely chopped)
1 cup glutinous rice (rinsed, soaked in water overnight)
goji berries (rinsed)

METHOD

1. Add 1/2 tsp of salt, 1/2 tsp of sugar, 1 tsp caltrop starch and the soaking water for American ginseng to the finely chopped pork. Mix well. Then add finely chopped American ginseng and water chestnuts. Stir well. Squeeze with your hands into pork balls.
2. Roll the pork balls in the glutinous rice to coat evenly. Garnish with 1 goji berry on top.
3. Steam over high heat for 10 minutes. Serve.

GIGI NOTES

For this recipe, do not add any spring onion as it is too overpowering and would cover up the flavours of the American ginseng.
Slice the pork first before rinsing. That would remove the gamey taste of the pork.
Soak the glutinous rice in water until it is saturated. Otherwise, the rice grains won't be soft enough after steamed.

Zha Jiang Pork Sauce

INGREDIENTS
900 g pork shoulder butt (shredded)
8 shallots (shredded)
2 tbsp cooking oil

SEASONING
8 tbsp ketchup
2 tbsp chilli sauce
2 tbsp sugar

METHOD
1. Stir-fry the shallot in oil until fragrant. Set aside.
2. In the same wok, stir-fry shredded pork for 5 minutes. The pork would give some juices during the process. Cook for about 3 minutes until the juices turn into oil. Put the shallot back in. Add seasoning and toss well. Cook for 5 more minutes.

GIGI NOTES

This pork sauce tastes even better if you let it sit overnight before serving. You may feel free to adjust the amounts of seasoning used according to your preference.
This pork sauce works magic on steamed rice or when stirred with blanched noodles. The photo here is Zha Jiang pork sauce on egg noodles.

Braised Pork Belly in Brown Sauce

INGREDIENTS

2 big strips pork belly (about 600 g, chopped into chunks, blanched in boiling water)
1 bird's eye chilli
1 tbsp rock sugar (crushed)
1 tbsp dark soy sauce
4 tbsp oyster sauce
1 1/2 cups Shaoxing wine
1/2 cup water
1 tbsp Zhenjiang vinegar (added at last)

METHOD

1. Stir-fry bird's eye chilli and rock sugar in 2 tbsp of oil until sugar dissolves. Put in the pork. Fry and toss until the pork is lightly browned.
2. Add dark soy sauce, oyster sauce, Shaoxing wine and water. Cook for 60 minutes until the sauce thickens. Add vinegar. Toss and serve.

GIGI NOTES

Blanching the pork helps remove its gamey taste, while making the pork less greasy.
Pork belly with alternate layers of fat and lean meat is the perfect cut for this recipe. The fat will be cooked till it melts in your mouth, without being greasy at all.

Braised Salted Pork Ribs with Bitter Melon and Soybeans in Casserole

INGREDIENTS

300 g pork ribs (marinated with 1 tsp of salt overnight)
2 bitter melons (de-seeded, cut into wedges)
1 cup soybeans (soaked in water overnight)
2 slices ginger

METHOD

1. Rinse off the salt on the pork ribs.
2. Heat a casserole pot. Add oil and stir-fry ginger until fragrant. Add pork ribs and soybeans. Pour in 2 cups of boiling water. Cook for 30 minutes until the pork ribs are tender.
3. When the sauce reduces to only a little, add bitter melon and cook for 2 minutes. Alternatively, if you prefer your bitter melon to be softer, cook for 5 minutes. Serve.

GIGI NOTES

After salted overnight, the meat on the ribs are firmer with an intense meaty flavour. It is best used for stews.
The soybeans are tender after soaking overnight in water. They will turn mushy more quickly that way.
I add boiling water to the pork ribs and soybeans so that the water won't reduce the temperature of the casserole and keep the heat while cooking.
Allrounder is an eco-conscious chopping board – recycled paper is stacked together, compressed, and bound by food-grade resin to create this lightweight and stable kitchenware. It is non-porous on the surface and stain-resistant while being free from heavy metal, volatile or semi-volatile organic compounds. The scratches and dents made by knife cuts it tend to be shallower so that germs and bacteria are less likely to hide on the chopping board.

Braised "Lion Head" Pork Balls

INGREDIENTS
600 g ground pork
1 head Napa cabbage (about 1.2 kg, cut into thick strips)
2 slices ginger

Marinade
1 tsp salt
1 tsp cooking wine
2 tsp sesame oil
2 tsp cornstarch
ground white pepper
2 tsp finely chopped spring onion
1 tsp diced ginger

SEARING GLAZE (MIXED WELL)
1 tbsp dark soy sauce
1 tbsp water
1 tbsp cornstarch

SEASONING
1 tsp salt
1 tsp sugar
1 tbsp dark soy sauce

METHOD
1. Add marinade to the ground pork and stir well. Lift it off the bowl and slap it back forcefully a few times. Then divide ground pork into four equal portions and shape each into a ball. Brush on the searing glaze and fry in a casserole pot with some oil until both sides golden. Carefully set aside the pork balls.
2. In the same pot, stir-fry ginger and Napa cabbage in the remaining oil. Add 1 cup of boiling water. Bring to the boil and put the pork balls back in and add seasoning. Cook over low heat for 30 minutes. Serve.

GIGI NOTES

The searing glaze acts like a coating on the skin of the pork balls. It forms a crust outside to seal in the juices, so that the pork balls are less likely to fall apart while still being succulent and moist inside.
This is a great dish to go with steamed rice – the cabbage would suck up the juices from the meat while the pork balls are juicy and flavourful.

Braised Pork Trotter Plum-Scented Wine Sauce

INGREDIENTS
1 pork trotter (chopped into chunks, blanched in boiling water)
150 g large peanuts

AROMATIC
4 slices ginger
4 cloves garlic (crushed)
4 shallots (finely chopped)
1 bird's eye chilli
1 tbsp rock sugar (crushed)

SEASONING
4 pickled sour plums
6 dried liquorice plums
2 tbsp Shaoxing wine
1 tbsp dark soy sauce

METHOD
1. Stir-fry aromatics in oil until fragrant. Put in the pork trotters and peanuts. Add seasoning. Then pour in enough water to cover all ingredients.
2. Bring to the boil over high heat. Turn to medium-low heat and cook for 90 minutes. Turn off the heat and leave it with the lid on for 30 minutes. Serve.

GIGI NOTES

Aromatics such as garlic, red chilli and shallots form the backbone of this dish's taste profile. They also impart much aroma.

Fried Beef Tenderloin in Sweet Sauce

INGREDIENTS

300 g beef tenderloin (sliced; marinated for 30 minutes)
4 tbsp sesame oil
2 baby cucumbers (grated into slices; mixed with 1/2 tsp of salt and marinated for 2 hours; squeezed dry; arranged on the rim of a serving plate)

SAUCE

2 tbsp sweet soybean paste
1 tbsp sugar
1 tbsp oil

MARINADE FOR BEEF

2 tbsp water
2 tbsp egg white
1 tbsp fish sauce
2 tsp caltrop starch

THICKENING GLAZE

1 tbsp Zhenjiang vinegar
1 tbsp dark soy sauce
1 tbsp Shaoxing wine
2 tsp ginger juice
1 tsp caltrop starch

METHOD

1. Heat sesame oil in a wok over medium heat. Fry the beef tenderloin until half-cooked. Set aside.
2. Cook the sauce ingredients over medium heat until sugar dissolves. Put the beef back in. Toss. Stir in the thickening glaze. Toss well and serve. Arrange cucumber slices on the side of the plate.

GIGI NOTES

This is a traditional Muslim recipe from Beijing. The original recipe calls for lamb. But some may find lamb too gamey. That's why I use beef instead.
There may be sinews on the beef tenderloin. Make sure you remove them.
I add egg white to the beef marinade as it helps keep the beef juicy and tender.

Fish Head Casserole with Lotus Roots

INGREDIENTS

1 head of bighead carp (including part of the body, cut into 4 big chunks, sprinkled with salt and pepper, rubbed evenly)
4 slices ginger

ACCOMPANIMENTS

1 small segment lotus root (peeled and sliced thinly)
1/2 onion (cut into chunks)
1 sprig Chinese celery (cut into short lengths)
8 cloves garlic (crushed)

SAUCE (MIXED WELL)

2 tbsp chilli bean sauce
2 tbsp sweet soybean paste
2 tbsp cooking wine
1 tbsp grated ginger
1 tbsp finely chopped spring onion
1/2 tsp salt
1 tsp sugar

METHOD

1. Heat a casserole and add oil. Stir-fry ginger until fragrant. Fry the fish head until all sides golden. Set aside.
2. Stir-fry all accompaniments in the remaining oil. Put the fish head on top. Pour in the sauce. Add 1/2 cup of boiling hot water. Cook for 10 minutes. Serve.

GIGI NOTES

Stir-fry the ginger before searing the fish head. The ginger helps remove the fishy taste.
As it takes time to suck on the fish head, I prefer serving this dish in a casserole pot to keep the fish warm for longer.

Fried Oyster Fritter

INGREDIENTS

300 g baby oysters (shelled, rubbed with caltrop starch, rinsed, wipe dry and diced. If you're using shucked oysters in a tub, finely dice them.)
2 shallots (finely chopped, stir-fried in oil till fragrant)
1 sprig coriander (finely chopped)
2 duck eggs (whisked)
3 tbsp fine sweet potato starch
1 tsp cornstarch
ground white pepper
1 tsp fish sauce
1/2 tsp oil

METHOD

1. Put all ingredients (except diced oysters and shallots) into a mixing bowl. Mix well.
2. Add oysters and shallots. Mix again. Pour the mixture into a heat pan with hot oil. Fry over low heat until both sides golden.

GIGI NOTES

Baby oysters come with many broken shells and there may be dirt and sand hidden in the folds. Thus, it's advisable to add caltrop starch to them and rub well to remove the dirt and rinse off the broken shells. You'll have the peace of mind that way.
I dice the baby oysters to make them cook more quickly. Alternatively, you may blanch them in boiling water until half cooked before dicing them.
For this recipe, I use fine sweet potato starch because it combine with water into a batter more easily. Adding sweet potato starch to the batter gives the fritter a chewy texture on the inside, while the crust will be crispy.
Oyster omelettes and fritters usually call for duck eggs, because of their intense yellow colour and intense eggy flavours.

Crispy Fried Eel Strips Covered in Cornflakes

INGREDIENTS

1 kabayaki eel (cut into thick strips)
coarse sweet potato starch
1 cup cornflakes (crushed finely)

SAUCE (MIXED WELL)

4 tbsp mayonnaise
1 tbsp condensed milk
1 tbsp wasabi (Japanese horseradish)

METHOD

1. Coat the eel strips in sweet potato starch. Deep-fry in oil until golden.
2. Then coat the eel strips evenly in the sauce. Coat them in crushed cornflakes. Serve.

GIGI NOTES

You may also use Rice Krispies or puffed rice instead of cornflakes.
Sweet potato starch comes in fine and coarse varieties. The coarse one is commonly used as a crust for deep-frying. Its bigger granules mean the crust would stay crispy for longer. Fine sweet potato starch is usually used to make oyster fritters.
This snack is a great match with an ice-cold beer.

Dry-Fried Curry Crabs

INGREDIENTS

2 crabs (scrubbed thoroughly with a toothbrush; dressed and chopped into pieces; coated lightly in caltrop starch; blanched in hot oil)

1/2 onion
2 shallots
2 cloves garlic
1 tbsp butter
1 tbsp cooking oil
2 tbsp homemade curry paste* (method in the following)
1/2 green bell pepper (cut into pieces, added at last)
1/2 red bell pepper (cut into pieces, added at last)
cooking wine

SEASONING

1 tbsp light soy sauce
1 tbsp coconut milk
1 small can evaporated milk

METHOD

1. Heat a wok and put in butter and cooking oil. Stir-fry onion, shallot and garlic until fragrant. Add 2 tbsp of curry paste.
2. Put the crabs into the curry paste. Sizzle with wine. Toss well. Add the seasoning. Put in bell peppers. Toss and serve.

**Curry Paste*

INGREDIENTS

2 onions (finely chopped)
4 cloves garlic (crushed)
4 slices ginger (crushed)
2 red chillies (finely chopped)
2 sprigs coriander
1 tsp turmeric
1 tsp ground cinnamon

METHOD

Put all ingredients into a blender or food processor. Puree. Heat a wok and add oil. Stir-fry the pureed mixture over low heat until fragrant. Add the liquids and seasoning. Bring to the boil.

LIQUIDS AND SEASONING

1 large can coconut milk
1/2 cup water
1 tsp salt
1 tsp ground black pepper

GIGI NOTES

You can make your own curry paste from scratch with this recipe. Or, feel free to get store-bought curry paste in a bottle. Just do whatever convenient to you.

Mung Bean Noodles with Mandarin Fish

INGREDIENTS

1 mandarin fish (de-boned and sliced, marinated for 12 hours)
2 sheets mung bean noodles (soaked in water till soft, torn into pieces)
1 cup stock

ACCOMPANIMENTS

1/2 green bell pepper (cut into chunks)
1/2 red bell pepper (cut into chunks)
2 slices ginger

MARINADE FOR FISH

2 tsp cooking wine
2 slices ginger (crushed)
2 sprigs white part of spring onion (crushed)
ground white pepper
1 tbsp caltrop starch

SEASONING (MIXED WELL)

3 tbsp ground soybean paste
2 tbsp sweet soybean paste
2 tbsp cooking wine
1 tsp dark soy sauce
1 tsp sugar

METHOD

1. Heat a wok and add oil. Stir-fry ginger over low heat until fragrant. Add green and red bell pepper. Toss a few times. Add seasoning and cook briefly. Set aside the bell pepper and ginger.
2. Pour stock into the sauce. Bring to the boil. Put in the fish and the mung bean noodles.
3. Cook until fish is done. Put the bell pepper and ginger back in. Serve.

GIGI NOTES

I used dry mung bean sheet noodles in the photo. They have to be rehydrated before use. You may also get fresh ones in the market. You can use those right away.
Besides using mandarin fish, you may also use grass carp fillet cut in butterflied slices instead.

Seared Grass Carp Fillet in Sichuanese Sauce

INGREDIENTS

1 grass carp fillet (marinated with salt and ground white pepper)

ACCOMPANIMENTS

4 slices ginger
4 cloves garlic
150 g soybean sprouts (with roots cut off)
1 half-ripened papaya (peeled, de-seeded, cut into chunks)
38 g dried kelp (soaked in water till soft; drained and cut into short lengths)
1 cube tofu (cut into chunks)

SEASONING

1 tbsp chilli bean sauce
1 tbsp sugar
1 tbsp light soy sauce
2 cups boiling water (added half at a time)

GARNISHES

1 sprig leeks (cut into short lengths)
1 red chilli (sliced)
1 sprig coriander (finely chopped)

METHOD

1. Heat a wok and add oil. Heat up the oil and put in the fish. Fry until both sides golden. Set aside.
2. In another wok, stir-fry ginger and garlic till fragrant. Then put in the rest of the accompaniments in the order listed above. Add seasoning and 1 cup of boiling water. Cover the lid and cook for a few minutes.
3. Put the fish back in. Add 1 more cup of boiling water. Bring to the boil and add the garnishes. Serve.

GIGI NOTES

When you fry the fish, make sure the wok and the oil are hot enough. The key is to brown the fish and fix its shape within a short period of time. That's how you keep it in one piece without falling apart.
The garnishes are there to add colours and aromas. You don't need to cook them through.

Roast Turkey in Chinese Style

INGREDIENTS
1 turkey (about 3.6 kg, marinated overnight)
2-3 glutinous rice dumplings in bamboo leaves (mashed)

MIREPOIX AND HERBS
1 small onion (chopped finely)
2 stems celery (chopped finely)
1/2 carrot (chopped finely)
2 bay leaves

MARINADE (DEPENDING ON THE SIZE OF THE CHICKEN)
2 tsp salt
2 tbsp dark soy sauce
2 tsp ground sand ginger
2 tbsp Shaoxing wine

METHOD
1. Stuff the turkey with the glutinous rice dumplings.
2. Preheat an oven to 230°C.
3. Put a grilling rack inside a big baking tray. Arrange mirepoix and herbs over the rack. Pour in a cup of boiling water. Put the turkey over the mirepoix with the breast facing up.
4. Bake over high heat at 230°C for 10 to 15 minutes first until the breast is lightly browned. Then cover the turkey in aluminium foil so that the skin won't brown too quickly. Turn the oven down to 170°C. For every pound of turkey (i.e. 450 g), bake for 12 minutes. (For example, a turkey weighing 8 pounds or 3.6 kg needs to be baked for 96 minutes to cook through.)
5. Turn off the oven and leave the turkey in the oven for 30 minutes for it to pick up the residual heat.

*Note: Turkey was not available when we were shooting for this book. Thus, chicken is used instead. The method is the same if you are using a turkey.

GIGI NOTES

Note that the initial 10- to 15-minute baking time over high heat, and the 30-minute resting time at last while the oven cools down do not count towards the baking time proper (i.e. 12 minutes per pound at 170°C).
If you don't like the turkey to be too pale in colour, you may brush straw mushroom dark soy sauce on the skin.
The amount of marinade depends on the size of the turkey. You may use the amounts listed here as a starting point.
I put the turkey over a bed of mirepoix and bay leaves so that their fragrance and sweetness will permeate through the turkey. The turkey is also less likely to stick to the baking tray.

Crispy Fried Chicken

INGREDIENTS

3 boneless chicken thighs (chopped into pieces, marinated overnight)
4 sprigs Thai basil (use the leaves only; deep-fried till crispy; as garnish)
1/2 cup coarse sweet potato starch
five-spice salt (served as a dip)

MARINADE

1 egg
2 tsp sugar
1 tbsp Shaoxing wine
1 tsp five-spice powder
2 tsp grated garlic
1 tbsp light soy sauce

METHOD

1. Sprinkle coarse sweet potato starch on the marinated chicken thighs. Deep-fry in hot oil at 180°C until cooked through. Drain and set aside.
2. Heat the oil again. Deep-fry the chicken thighs once more over medium heat. Drain. Serve with five-spice salt on the side.

GIGI NOTES

Make sure you wipe the Thai basil dry before deep-frying it. Otherwise, the hot oil may splatter.
After you coat the chicken pieces in sweet potato starch, it's advisable to deep-fry them right away. Otherwise, the crust won't be as crispy if the starch turns damp after picking up the moisture from the chicken.
I prefer using coarse sweet potato starch. It gives the crust a lovely crunch.
This fried chicken tastes like Taiwanese fried chicken from night markets.

Braised Chicken with Cabbage

INGREDIENTS

1 chicken
38 g dried conches (see preparations on p.183)
1 cabbage (cut into thirds)
1 cup rice wine
2 cubes rock sugar
4 cloves garlic
1 sprig Peking scallion (cut into short lengths)
4 cups water
1/2 cup light soy sauce
4 shallots

METHOD

1. Line the bottom of a pot with one-third of a cabbage. Put the whole chicken on top.
2. Put the remaining two-thirds of the cabbage around the chicken. Put in all remaining ingredients. Bring to the boil and turn to low heat. Simmer for 3 hours. Serve.

GIGI NOTES

If you have Shanghainese or Korean rice cakes at home, blanch them in boiling water first. Then put them in with the braised chicken. Cook till tender and serve.

Stir-Fried Chicken and Abalone with Thai Basil

INGREDIENTS

4 boneless chicken thighs (cut into chunks, marinated)
1 can abalone (double-boiled in can for 2 hours) (cut into pieces)
2 cubes rock sugar
4 slices ginger
1 red chilli
2 tsp sesame oil
1 tbsp fish sauce
1 onion (cut into chunks)
4 shallots (gently crushed)
4 sprigs Thai basil (use leaves only)
2 tbsp aged vinegar

MARINADE FOR CHICKEN

1/2 tsp salt
1 tbsp rice wine
1 tsp sugar
1/2 tsp ground black pepper

METHOD

1. Fry the chicken pieces in 2 tbsp of oil until medium-well done. Set aside.
2. In the same wok, add 2 tbsp of sesame oil and stir-fry onion, shallots, ginger and rock sugar until fragrant. Put the chicken back in and add the abalone. Sprinkle with aged vinegar and fish sauce. Toss for 2 minutes. Add Thai basil leaves. Toss quickly. Serve.

GIGI NOTES

I put in the Thai basil leaves right before I dish up to retain their aromas.
For this recipe, I double-boil the whole can of abalone in water. Boil enough water in a pot to cover the whole can. Then cook for 2 hours over low heat. Drain and let cool. Then open the can and use the abalone in the recipe. The slow cooking process makes the abalone tender and soft.
As the chicken and abalone require different cooking times, make sure you cook them separately at first, before braising them together for the flavours to mingle.

Braised White Radish with Preserved Pork Belly in Mala Sauce

INGREDIENTS

1/2 strip Cantonese preserved pork belly (blanched in boiling water; sliced diagonally)
1 white radish (peeled, cut into thick strips)
1 sprig Peking scallion (sliced diagonally)
8 cloves garlic
1 red chilli (sliced diagonally)
1 tbsp Sichuan peppercorns (To make Sichuan pepper oil, put oil into a cold wok. Add Sichaun peppercorns and half of the red chilli. Fry them until the oil turns red and you can still the Sichuan peppercorns. Strain the oil and set aside.)

SEASONING (MIXED WELL)

1 tbsp oyster sauce
1 tsp dark soy sauce
1/8 raw cane sugar slab (finely chopped)
ground white pepper
1 cup boiling water

METHOD

1. Stir-fry garlic and the remaining red chilli in the Sichuan pepper oil until fragrant. Add radish and preserved pork belly. Toss again.
2. Add seasoning. Turn to high heat to reduce the sauce. Add Peking scallion. Toss well. Serve.

GIGI NOTES

I blanch the preserved pork belly to remove some of the grease. It is also easily to slice after blanched.
I discard the Sichuan peppercorns after making the Sichuan pepper oil. When you eat it, you don't want to bite on to a hard peppercorn that releases a strong flavour in your mouth.
I always put oil and Sichuan peppercorns in cold wok and start heating it up slowly. That gives time for the Sichuan peppercorns to release their fragrance. If you heat the wok and oil first and put them in, they tend to burn and char easily and become bitter in taste.

Braised White Radish

INGREDIENTS

2 white radishes (peeled, cut into random wedges)

2 cups stock   2 slices ginger   1/4 slab raw cane sugar (crushed)

METHOD

1. Heat a wok and add oil. Stir-fry ginger and cane sugar until fragrant. Put in the white radishes and toss well. Pour in the stock. Cook over low heat for 1 hour.
2. Arrange the radishes on a serving dish. Stir in caltrop starch slurry to thicken the sauce. Drizzle this sauce over the radishes. Serve.

GIGI NOTES

I usually cut white radishes into random wedges while rolling them on the chopping board. Random wedge shape works especially well for braised recipe like this one.

When you shop for white radishes, pick those that feel heavy in your hand. That means they are juicy and tender. For such white radishes, you don't need to peel them deeply as their skin is thin and not very fibrous.

Alternatively, you can put all ingredients into a pressure cooker and cook them for 30 minutes to save time.

Not only does the cane sugar flavour the radishes, but also soften them. Thus, recipes of braised beef briskets or beef shin usually call for a cube of rock sugar.

This book is scheduled to be released in summer. When you make this dish, feel free to get one more radish and make yourself an appetizing pickled radish skin. The ingredients include white radish, sweet black rice vinegar, white vinegar, Shanxi aged vinegar, Maggi's seasoning, white sugar and bird's eye chillies. Do not peel the radish and just slice it thickly. Add 1 tbsp of salt and mix well. Marinate the radish for 30 minutes. Then squeeze dry. Put the radish, red chilli and all remaining ingredients into a bowl. Leave it to soak for 12 hours. Serve.

Knives from the Swiss Modern series of Victorinox are made with high-strength stainless steel that is resistant to abrasion. Their sharp blades make slicing food easy and effortless. The cut is always smooth and you can handle every cutting task with finesse and precision. Cooking has never been easier.

Braised Baby Napa Cabbages with Pork Skin

INGREDIENTS

3 baby Napa cabbages (cut in half, and cut into thick strips)
8 dried shiitake mushrooms (see p.183 for preparations, then cut into thick strips)
4 slices ginger (crushed)
1 large piece puffed pork skin (soaked in water till soft; cut into pieces; blanched in water with ginger, spring onion and vinegar; drained and squeezed dry)

SEASONING

1 tbsp oyster sauce
1 tsp dark soy sauce
1 tsp ground white pepper
1 tsp sesame oil

METHOD

1. Heat a wok and add oil. Stir-fry ginger, cabbages, pork skin and shiitake mushrooms briefly. Add 1 cup of hot water.
2. Cook over medium-high heat for 15 minutes. Add seasoning and toss well. Serve.

GIGI NOTES

Some frozen deli stores actually carry rehydrated puffed pork skin. Just blanch it and use it right away.
You can get dried puffed pork skin in grocery stores. You need to soak it in cold water until soft. Then cut into pieces and blanch it in boiling water with ginger, spring onion and vinegar. That helps remove the excess grease on the pork skin and remove the stale smell.

Stir-Fried Cabbage in Spicy Vinegar Sauce

INGREDIENTS

1 small cabbage (torn into large leaves)
1 tbsp Sichuan peppercorns
8 dried bird's eye chillies (cut finely with scissors)
2 tbsp finely diced ginger

SEASONING

2 tbsp aged vinegar
1 tbsp light soy sauce
2 tbsp sugar
1/2 tsp salt
caltrop starch slurry (2 tbsp caltrop starch mixed with 1/2 cup water)

METHOD

1. Heat a wok and add 2 tbsp of oil. Stir-fry Sichuan peppercorns and half of the dried chillies over low heat until the oil is fragrant. Remove peppercorns and chillies. Keep the scented oil in the wok.
2. In the same wok, stir-fry the remaining dried chillies and diced ginger until fragrant. Turn to high heat and put in the cabbage. Add seasoning. Keep tossing until the cabbage is just cooked and still crunchy. Serve.

GIGI NOTES

You may use Napa cabbage instead of white cabbage, but if you use Napa cabbage, you should cook it till it wilts and turns soft. On the other hand, white cabbage should still be crunchy and crisp after cooked.
If you use Napa cabbage, cut it into thick strips before stir-frying. It picks up heat and cooks more quickly that way.

Stir-Fried Diced Chillies with Lotus Roots and Fermented Black Beans

INGREDIENTS

1 segment lotus root (diced, soaked in water to wash off the starch, rinsed repeatedly, blanched in boiling water with a pinch of sugar)
2 long green chillies (diced)
2 long red chillies (diced)

AROMATICS

4 slices ginger (diced finely)
4 cloves garlic (finely chopped)
2 tbsp fermented black beans (coarsely chopped; with sugar and oil added; mixed well)

SEASONING

2 tbsp light soy sauce
1 tbsp sugar
1 tbsp vinegar

METHOD

1. Stir-fry diced lotus root in a wok. Add aromatics and toss until the lotus root picks up the flavours.
2. Add red and green chillies. Toss briefly. Add seasoning and toss to mix well. Serve.

GIGI NOTES

You may also add ground pork if you want.
If you don't wash off the starch from the lotus root, it would stick together when stir-fried. On the other hand, if you are making a stew or a braised dish with lotus root, do not wash off the starch.
When you blanch the lotus root in water, add a pinch of sugar to remove the grassy taste.

Stir-Fried Broccoli and Cauliflower

INGREDIENTS

1/2 head cauliflower (cut into florets)
1 head broccoli (cut into florets)
3 cloves garlic (grated)
1 dried red chilli (finely chopped)
1 tsp salt
1 tsp sugar
sesame oil

METHOD

1. Blanch the cauliflower and broccoli in boiling water with 1 tbsp of oil and 1 tsp of salt for 2 minutes. Drain.
2. Heat a wok and add oil. Stir-fry garlic and red chilli until fragrant, but don't burn them. Turn to high heat. Put in the cauliflower and broccoli. Season with salt and sugar. Drizzle with sesame oil. Toss well and serve.

GIGI NOTES

After cutting the broccoli into florets, soak them in a bowl of water for 15 minutes with 1 to 2 tsp of salt added. The bugs hidden between the tiny buds would come out and float on in the water. Then rinse them once more and they are ready to cook.
Cauliflower and broccoli are firm in texture. Blanching them first before stir-frying helps pre-cook them slightly and shortens the time required for stir-frying.

Sweet Corn Kernels in Salted Egg Yolk Sauce

INGREDIENTS

3 ears sweet corn (with kernels sliced off)

8 salted egg yolks (steamed till done; mashed while still hot)

1/2 red bell pepper (diced)

1/2 green bell pepper (diced)

1 tsp butter

1 tsp oil

1 tbsp Shaoxing wine

1/2 tsp salt

1/2 tsp sugar

ground black pepper

METHOD

1. Heat a wok and add oil. Turn to low heat. Melt the butter. Put in salted egg yolks and keep tossing until they bubble. Sizzle with Shaoxing wine.
2. Put in the sweet corn kernels. Turn to high heat. Add salt and sugar. Toss quickly for 2 minutes. Add bell peppers and toss again. Sprinkle with ground black pepper. Toss and serve.

GIGI NOTES

For this recipe, you must use fresh sweet corn on the cob and cut off the kernels yourself. Do not use canned or frozen corn kernels because they tend to give much water when fried and are not good in texture. Only fresh sweet corn kernels would turn into vesicles that burst with sweet juices with your every bite. This is the ultimate pleasure of this dish.

Butter and salted egg yolks are the perfect matches with each other. The salted egg yolks are fragrant only when fried in butter. But butter tends to burn easily. Thus, I always add a dash of cooking oil first, before putting in the butter. That would stop it from burning.

You can freeze the leftover sweet corn cobs for later use, such as making soups. Don't be wasteful.

Stir-Fried Celtuce in Chilli Black Bean Sauce

INGREDIENTS

1 head celtuce (peeled, tough veins removed; cut into thick strips, marinated with 1 tsp of salt)
1 tbsp fermented black beans (soaked in hot water briefly)
1 tbsp grated garlic
1 tbsp diced ginger
1 red chilli
1 tbsp Shaoxing wine
1 tsp sugar
sesame oil

METHOD

1. Heat a wok and add oil. Stir-fry ginger, garlic and fermented black beans until fragrant. Sizzle with Shaoxing wine.
2. Put in the celtuce and chilli. Toss quickly. Add sugar and sesame oil. Toss again. Serve.

GIGI NOTES

I marinate the celtuce with 1 tsp of salt to draw the moisture out of the celtuce and to make it crisper in texture.
I soak the fermented black beans in hot water to soften them slightly. That helps release their flavours more readily when stir-fried.

Fruit Salad with Chicken Breast

INGREDIENTS

2 boneless chicken breasts (sliced thinly diagonally, marinated for 2 hours)
1 cucumber (grated into long strips)
1/2 pineapple (peeled, cut into strips)
8 strawberries (quartered)
1 box blueberries

MARINADE FOR CHICKEN

1 pinch salt
1/2 tsp lemon juice
1 tsp caltrop starch
1 tsp olive oil

DRESSING (MIXED WELL)

1/2 tsp salt
1 tsp sugar
1 tsp lemon juice
10 mint leaves (crushed)

METHOD

1. Heat a pot of water until bubbles appears on the bottom. Add 1 tsp of salt and 1 tbsp of oil. Blanch the chicken slice by slice for 10 seconds each. Set aside.
2. Mix all ingredients. Toss in the dressing. Serve.

GIGI NOTES

For this recipe, the chicken should be blanched until just cooked. It would stay juicy and tender that way.

Healthful Breakfast Oatmeal Cookies

INGREDIENTS

115 g butter (softened at room temperature, or heated for 30 seconds in a microwave oven)
1/2 cup light brown sugar
1 large egg (whisked)
3/4 cup flour
1/2 tsp baking powder
1/2 tsp ground cinnamon
a pinch salt
(the four ingredients above: sieved together)
1 1/2 cups instant rolled oats
1/2 cup raisins

METHOD

1. In a mixing bowl, stir together butter, light brown sugar and egg. Sieve in the dry ingredients. Mix well. Add rolled oats and raisins. Stir well into dough.
2. Shape the dough into small round cookies. Bake in a preheated oven at 170°C for 15 minutes. Flip them upside down and bake for 10 more minutes. Serve.

GIGI NOTES

You may add coarsely chopped walnuts to the cookies, for extra nuttiness and nutrients.

After you leave the cookies to cool completely, store them in an airtight container.

I serve these cookies at breakfast and I would make a whole week's worth of supply at one time. You may serve them with milk, or crush them and sprinkle over yoghurt.

This recipe is extremely easy with no skills required. It is completely fail-proof.

Baked Cantonese Pork Sausage Pancakes

INGREDIENTS A (*SIEVED TOGETHER, EXCEPT BUTTER)

57 g chilled butter (diced)
1 tbsp sugar
1/4 tsp baking soda
1 tsp baking powder
1 cup plain flour
1/4 tsp salt
1/2 tsp garlic powder

INGREDIENTS B (*MIXED WELL, THEY SHOULD ADD UP TO 1 CUP)

190 ml milk
1 egg (whisked)

INGREDIENTS C

1 Cantonese preserved pork sausage (finely chopped)
1 sprig spring onion (finely chopped)
1 cup grated cheeses of your choice

METHOD

1. Preheat an oven to 180°C.
2. Put all ingredients A into a mixing bowl. Mix well. Beat with an electric mixer for about 1 minute until butter melts.
3. Put in ingredients B. Keep stirring until well combined. Lastly add ingredients C. Use a rubber spatula to stir until it becomes a smooth thick paste.
4. Line a baking tray with parchment paper. Ladle batter on it to make round pancakes. Bake in the preheated oven for 20 to 25 minutes.

GIGI NOTES

This is a fusion recipe that blends Chinese and Western ingredients and cooking techniques. The rich aroma of Cantonese pork sausage works unexpectedly well with cheese and spring onion. It is a great afternoon snack to go with teas.
The key step of this recipe is dicing the butter. Otherwise, it won't melt when beaten with an electric mixer.
When you ladle the batter on the baking tray, make sure you leave enough space between the pancakes. They would expand in the baking process and may stick together in one piece if they don't have enough space around them.

Coconut Panna Cotta

INGREDIENTS

3/4 cup milk
1/3 cup evaporated milk
2/3 cup coconut milk
1/4 cup sugar
2 tsp gelatine powder (mixed with 4 tbsp of warm water)

METHOD

1. Put all ingredients (except gelatine powder) into a small pot. Cook over low heat until sugar dissolves.
2. Pour in the gelatine solution and stir well. Turn off the heat. Let cool and pour into cups or a mould. Refrigerate until set. Serve.

GIGI NOTES

I dissolve the gelatine powder in warm water first before adding it to the coconut milk mixture. If you put dry gelatine powder in directly, the gelatine may clump together and the panna cotta may not set properly.
Do not boil the coconut milk mixture vigorously. It should only come to a gentle simmer as long as the sugar dissolves. If you boil it too vigorous, the panna cotta will not be silky smooth.

Pumpkin Pancakes with Red Bean Filling

INGREDIENTS

1/2 Japanese pumpkin (about 600 g, de-seeded, steamed for 15 minutes, peeled, mashed while still hot)
300 g glutinous rice flour
3 tbsp sugar
red bean filling (store-bought)
sesames

METHOD

1. In a mixing bowl, mix together mashed pumpkin, sugar and glutinous rice flour. Knead into dough.
2. Roll a piece of dough into a ball. Then press to flatten it. Wrap in some red bean filling and fold the edges toward the centre to seal well. Pinch the seams. Then press it flat again. Brush some water on the side with no seams. Press it firmly on sesames to coat it evenly.
3. Heat a frying pan. Add oil. Fry the pancakes over low heat until both sides golden. Serve.

Candied Gingkoes with Osmanthus

INGREDIENTS

2 cups gingkoes (shelled and cored)
1/2 cup rock sugar
1 to 2 tbsp candied osmanthus

METHOD

1. Put gingkoes and rock sugar into a pot. Add water to cover. Bring to the boil over high heat. Turn to medium heat and stir continuously.
2. Cook until the gingkoes are tender and the liquid reduces. Stir in candied osmanthus. Serve.

GIGI NOTES

You can get candied osmanthus from shops specializing in Shanghainese grocery.
Do not eat too many gingkoes at one time. This sweet snack is supposed to be shared among a few friends. The gingkoes are chewy and sweet, with a hint of mild bitterness.

Banana Chiffon Cake

INGREDIENTS A

7 egg whites (beaten over medium-high speed for about 3 minutes until stiff)

INGREDIENTS B

3 bananas (mashed with a fork)
1/2 cup milk
1/2 cup vegetable oil
7 egg yolks
1/2 cup sugar

INGREDIENTS C

1 1/2 cups plain flour
1 tbsp baking powder
a pinch salt
(sieved together)

METHOD

1. Preheat an oven to 160°C (305°F).
2. Put ingredients B into a mixing bowl. Beat briefly to mix well. Slowly sieve in ingredients C and beat after each addition.
3. Put ingredients A into the flour mixture from step 2. Fold gently to mix well.
4. Transfer into a chiffon flute tin. Bake in the preheated oven for 1 hour. Leave it to cook in the tin upside down. Unmould and slice. Serve.

GIGI NOTES

For this recipe, use fully ripened bananas with black dots on the skin. They have strong fruity aromas with soft flesh and are perfect for baking.
Do not keep bananas in the fridge. Otherwise, their flesh would turn dark.
Just leave them on the counter and wrap the stems in cling film. That would stop them from ripening too quickly.
Use a chiffon flute tin as shown in the photo on p.94. And leave the cake to cool in the tin upside down. The cake hasn't fully set yet straight out of the oven. If you cool it the right way up, gravity would pull and the cake would collapse and shrink. The cake would end up dense instead of airy and fluffy.
Chiffon cakes do not need any rising agent such as baking soda. It rises because of the tiny air bubbles trapped in the meringue that expand when heated.

Fruit Jam Cake

INGREDIENTS A

225 g unsalted butter (softened at room temperature)
1/2 cup sugar
4 eggs (at room temperature)
4 tbsp fruit jam

INGREDIENTS B

1 1/2 cups plain flour
1 1/2 tsp baking powder
a pinch salt

METHOD

1. Preheat an oven to 170°C.
2. For ingredients A, put butter in a mixing bowl. Beat with an electric mixer over high speed for 1 minute. Add half of the sugar at one time. Beat well after each addition until sugar dissolves. Then crack in one egg at one time. Beat until well incorporated after each addition. Stir in the fruit jam at last.
3. For ingredients B, sieve all ingredients together. Add half of the mixture to the egg mixture from step 2 at one time. Beat over low speed after each addition until just mixed.
4. Pour the batter into a cake tin. Bake in the preheated oven for 40 minutes. Serve.

GIGI NOTES

You may use any fruit jam or fruit preserve of your choice. The photo shows one made with strawberry jam and you may serve it with fresh strawberries. If you use blueberry jam in the cake, you may also put in some fresh blueberries in the cake batter.
When you add eggs to the butter-sugar mixture, you should only put in one egg at a time. That would stop the eggs from separating with the butter. Otherwise, the butter would sink to the bottom, and the cake would not be of the same consistency throughout.

Classic Lasagne in Meat Sauce

INGREDIENTS

1 pack lasagne
450 g ground beef
225 g ground pork
assorted grated cheeses (added last)

SEASONING

1 tsp ground black pepper
1 tsp salt

ACCOMPANIMENTS

1 onion (chopped)
3 tomatoes (chopped)
1 can peeled tomatoes (about 400 g)
225 g button mushrooms (chopped)
2 bay leaves
1/2 slab raw cane sugar
2 tbsp tomato sauce
2 tbsp tomato paste

METHOD

1. Stir-fry the ground beef and pork in a dry wok until all juices dry out. Add seasoning and toss well. Set aside.
2. In another wok, melt butter and add some oil. Stir-fry onion until transparent. Add remaining accompaniments. Cook over low heat for 30 minutes.
3. Add ground beef and pork into the tomato sauce from step 2. Cook briefly.
4. In a baking tray, line the bottom with a layer of lasagne. Scoop a ladle of meat sauce on top. Then stack another layer of lasagne over and top with meat sauce. Repeat this step to build alternate layers of paste and sauce. The last layer should be meat sauce and generously sprinkle with grated cheeses on top.
5. Bake in a preheated oven at 190°C (375°F) for 40 to 50 minutes.

GIGI NOTES

Instead of using ground pork and beef, you can may also use raw chorizo or sausage links for this recipe. Just squeeze the meat out of the casing and use it in place of ground pork and beef. But as the sausage filling has been seasoned, it's advisable to taste the meat sauce first before seasoning it further with salt.

Braised Oil Noodles with Fried Prawns

INGREDIENTS

300 g oil noodles

300 g large marine prawns (with salt rubbed on them twice to draw out the moisture; wiped dry; marinated with a pinch of ground white pepper)

1 cup prawn stock

150 g French beans (with tough veins torn off; sliced diagonally; blanched in hot oil briefly; drained)

2 eggs (whisked; fried into omelette in a pan; finely shredded)

2 slices ginger (shredded)

2 cloves garlic

1/2 tsp salt

METHOD

1. Cut the prawns along the back deeply, but without cutting all the way through. Devein and rinse. Drain and wipe dry. Blanch in hot oil. Drain.
2. Stir-fry ginger and garlic until fragrant. Add the prawn stock and bring to the boil. Put in oil noodles and cook for 2 minutes. Add French beans, shredded omelette and salt. Cook for 1 minute.
3. Put the prawns back in. Toss well. Serve.

GIGI NOTES

Whenever I cook prawns and shrimps, I would collect the heads and shells and make prawn stock with them. The stock works in many recipes to impart umami and richness.

To make the prawn stock, stir-fry the prawn heads and shells in butter until fragrant. Add 1 cup of water and boil for 3 minutes until the stock looks red. Strain the stock and set aside the stock.

Besides flavours, a perfect dish should also look good. Thus, I always cut the prawns along the back and remove the veins before cooking them.

Lean Pork Soup with Moon Scallops and Goji Berries

INGREDIENTS

600 g lean pork (blanched in boiling water)
38 g dried moon scallops (rinsed)
38 g dried goji berries ⎤ soaked in water and rinsed
38 g Huai Shan (dried yam) ⎦
2 ears sweet corn (cut into quarters)
2 carrots (peeled and cut into pieces)
1 whole dried tangerine peel (soaked in water till soft)
3 candied dates
18 cups water

GIGI NOTES

This soup benefits the Liver, tonifies the Lungs, and improves eyesight.

METHOD

1. Boil the water in a pot. Put in all ingredients.
2. Boil over high heat for 10 minutes. Turn to medium heat and cook for 3 hours. Serve. If you use a pressure cooker, you only need to add 13 cups of water and cook for 1 hour.

Chicken Soup with Starfruits and Apples

INGREDIENTS

1 chicken (blanched in boiling water, skinned)
4 red delicious apples (cut into quarters, cored)
4 starfruits (cut into thirds, de-seeded)
2 tbsp sweet and bitter almonds
18 cups water

METHOD

Put all ingredients into a pot. Boil over high heat for 10 minutes. Turn to medium heat and cook for 3 hours.

GIGI NOTES

I prefer using red delicious apples in soups, because they are sweet and juicy, without the tartness that some other varieties carry.
Always skin the chicken before making soup with it. That would make sure the soup is flavourful, yet without the greasiness.

Dace Soup with Kudzu, Burdock and Cuttlefish

INGREDIENTS

1.2 kg kudzu (peeled, cut into pieces about the size of a mah-jong tile)
600 g burdock (with skin scraped off, sliced thickly)
38 g dried Tu Fu Ling
38 g small red beans
38 g hyacinth beans
38 g malted rice
38 g raw malted wheat
(the five ingredients above: soaked in water for 1 hour)
1 dace (fried in oil with a slice of ginger)
1 dried cuttlefish (with bone removed, soaked in water till soft, cut into large pieces)
5 slices ginger
3 candied dates
18 cups boiling water

METHOD

1. In a pot, fry the fish in a little oil with a slice of ginger until golden on both sides. Add boiling water. Put in the rest of the ingredients.
2. Boil over high heat for 10 minutes. Turn to medium heat and cook for 3 hours. Serve.

GIGI NOTES

To make soup, old kudzu with clear lines on the skin works best because of its strong flavours. I prefer cutting the kudzu into the size of a mah-jong tile for the flavours to infuse easily.
If you use a pressure cooker, you only need to add 13 cups of water and cook for 1 hour.
This soup benefits the Spleen, dissipates Fire, eases indigestion, neutralizes stomach acid, and get rid of bad breath.
When you eat this soup, beware of any stray bones. Or, you may put the fish in a muslin bag and tie it well before putting it into the soup. That would save you the risk of choking on bones.

Seven-Coloured Spicy and Sour Thick Soup

INGREDIENTS

5 cups water
1/2 white radish (grated into thin strips)
1 pack Konnyaku noodles (cut into short lengths)
1 tbsp shredded wood ear fungus (soaked in water till soft)
4 dried shiitake mushrooms (soaked in water till soft, shredded)
1/2 carrot (grated into thin strips)
60 g pork tenderloin (shredded, marinated with salt and ground white pepper)
1 sprig Chinese celery (cut into short lengths) (added last)

SEASONING

1 tsp salt
1 tbsp aged vinegar
ground white pepper
1 tsp sugar
1 tbsp dark soy sauce

THICKENING GLAZE (MIXED WELL)

2 tbsp caltrop starch
2 tbsp water

METHOD

1. Boil a pot of water. Put in all ingredients except pork and Chinese celery. Boil for 5 minutes. Skim off the froth on top.
2. Add shredded pork. Cook for 1 minute. Add seasoning and bring to the boil. Stir in the thickening glaze and cook till it thickens. Sprinkle with Chinese celery on top. Serve.

GIGI NOTES

The pork cooks very quickly as it has been shredded. I put it in the soup at last to keep it tender. For the same reason, the Chinese celery goes in the soup at last so as to keep it bright green and aromatic.

Honeydew Melon Soup with Sha Shen and Yu Zhu

INGREDIENTS

1 honeydew melon (about 900 g, rinsed, de-seeded, cut into chunks with skin on)
300 g lean pork (blanched in boiling water)
8 chicken feet (blanched in boiling water)
150 g dried conches (see p.183 for preparations)
3 dried scallops
38 g Sha Shen (soaked in water and rinsed)
38 g Yu Zhu (soaked in water and rinsed)
4 dried figs (rinsed, with a cut made at the centre)
15 cups water

METHOD

Put all ingredients into a pot. Boil over high heat for 10 minutes. Turn to medium heat and cook for 3 hours. Serve.

GIGI NOTES

I make a cut at the centre of each fig with scissors, so that the flavours would infuse more readily.
Adding chicken feet to this soup adds an extra dimension of richness. You can literally feel the collagen with every gulp.

Cream of Sweet Corn Soup with Bamboo Fungus

INGREDIENTS

6 strips dried bamboo fungus (soaked in water till soft; rinsed and cut into rings)
1 boneless chicken breast (diced finely, marinated with salt and pepper)

SOUP BASE

1 can chicken stock (about 250 ml)
1 pack cream-style sweet corn (about 415 g)
3 cups water

THICKENING GLAZE (MIXED WELL)

2 tbsp water chestnut starch
1/2 cup water

METHOD

1. Boil the soup base in a pot. Put in the bamboo fungus and cook briefly. Put in the chicken and bring to the boil. Stir in the thickening glaze and bring to the boil again.
2. Sprinkle with finely chopped spring onion and ground white pepper. Serve.

GIGI NOTES

If you want give this soup more chewy texture, add 1 cup of fresh sweet corn kernels.

Lean Pork Soup with American Ginseng and Shi Hu

INGREDIENTS

600 g lean pork (blanched in boiling water)
38 g American ginseng (soaked in water briefly)
38 g Shi Hu
38 g dried lily bulbs
38 g Hai Yu Zhu
(the above three: soaked in water and rinsed)
38 g dried figs (rinsed in water, with a cut made at the centre)
115 g sliced conch (see p.183 for preparations)
18 cups water

METHOD

Put all ingredients into a pot. Boil over high heat for 10 minutes. Turn to medium heat and cook for 3 hours. Serve.

GIGI NOTES

This soup moistens the Lungs and is especially good for those who always stay up late at night and those who smoke or consume alcohol frequently.
American ginseng strengthens Qi (vital energy) flow, regulates the Heart meridian, and enlivens the spirits. Lily bulbs nourishes the Yin, moistens the Lungs, clears the Heart meridian and calms the nerves. Shi Hu nourishes the Yin, beautifies the skin and stimulates the secretion of body fluids.

Fish Head Soup with Assorted Beans

INGREDIENTS

300 g pork shin (blanched in boiling water)
1 head of bighead carp (about 600 g, fried in oil with 1 slice ginger)
1/2 cup cashew nuts
1/3 cup peanuts
1/3 cup black-eyed beans
1/3 cup black beans
1/3 cup soybeans
1 piece dried tangerine peel
(the above: soaked in water for 30 minutes)
15 cups boiling water
4 red dates (de-seeded)

METHOD

1. Fry the fish head in a pot with a little oil and a slice of ginger until browned on both sides. Pour in 15 cups of boiling water. Then put in all remaining ingredients.
2. Boil over high heat for 10 minutes. Turn to medium heat and cook for 3 hours. Serve. If you use a pressure cooker, you only need to add 10 cups of water and cook it for 30 minutes.

GIGI NOTES

I fry the fish head with a slice of ginger to remove the fishy taste.
I add boiling water to the fish so that the temperature of the wok or pot won't drop drastically. That also helps shorten the cooking time.

Pork Tripe Soup with White Peppercorns and Monkey Head Mushrooms

INGREDIENTS

38 g dried monkey head mushrooms (soaked in water till soft)
38 g white peppercorns (cracked gently)
115 g white lotus seeds (soaked in water for 1 hour)
1 pork tripe (rinsed; fat trimmed off; blanched in boiling water)
75 g dried oysters (soaked in water till soft; rinsed to remove dirt)
5 slices ginger
18 cups water

METHOD

1. Put all ingredients into a pot. Cook over high heat for 10 minutes. Turn to medium heat and cook for 3 hours.
2. If you use a pressure cooker, you only need to add 13 cups of water and cook for 1 hour to make this soup.

GIGI NOTES

You can get rinsed and cleaned pork tripe from your butcher in traditional wet markets. But there are not many of them available. You should order one day ahead.
After soaking the dried oysters, try to clean it thoroughly. There may be broken shells sticking on the dried oysters. There may also be dirt and dust hidden in the gills.
White peppercorns warm the central organs, ease indigestion, dissipate phlegm and detoxify. Cracking the white peppercorns helps release their flavours more easily.
Monkey head mushrooms benefit the Stomach, ease Asthenia, and prevent stomach ulcers and tumours.
This soup whets the appetite, strengthens the Spleen, warms the Stomach, regulates Stomach functions, promotes Qi (vital energy) flow, and eases indigestion.
I use low pressure cooker from the WOLL brand for this recipe. It creates a stronger pressure inside the pot than regular cookware, thereby reduces braising time by half and cutting down fuel consumption by 70%. Not only does it save on fuel cost, but also reduces the impact on the environment. It is more convenient than pressure cookers as you may open the lid halfway through the cooking time to add ingredients. You don't need to wait for the pressure to release before serving the food either.

Silkie Chicken Soup with He Shou Wu and Conches

INGREDIENTS

1 Silkie chicken (dressed and blanched in boiling water)
38 g He Shou Wu
38 g Dang Gui head
38 g Tu Fu Ling
(rinsed)
38 g black beans
38 g large peanuts
(soaked in water for 2 hours)
38 g red dates (de-seeded)
38 g dried conches (see p.183 for preparations)
1 whole dried tangerine peel (soaked in water till soft)
5 slices ginger
18 bowls water

METHOD

Put all ingredients into a pot. Boil over high heat for 10 minutes. Then turn to medium heat and boil for 3 hours.

GIGI NOTES

Most of the herbal ingredients in this recipe are useful in tonifying the Liver and Kidneys, while regulating the blood chemistry. This soup helps prevent hair loss and grey hair.
You don't need to skin the chicken for this soup. The soup may turn out greasier, but it will also be more flavourful. I suggest straining the soup in a fat separator pot. That would remove the fat of the soup.

Apple Cream Soup

SOUP BASE

5 apples (quartered; cored)
2 celery stems (with tough veins torn off, finely chopped)
1 carrot (peeled, finely chopped)
3/4 onion (finely chopped)
2 slices bay leaves
4 cups water

INGREDIENTS

7 apples (peeled, cored, diced)
1 potato (peeled, diced)
1 litre whole-fat milk

GARNISHES

whipping cream
ground cinnamon

METHOD

1. To make the soup base, stir-fry mirepoix (i.e. celery, carrot and onion), bay leaves and apples until fragrant. Add water. Bring to the boil and turn to low heat. Cook for 1 hour. Strain the mixture and save the soup base only.
2. Bring the soup base from step 1 to the boil. Add 7 apples, 1 potato and 1 litre of milk. Bring to the boil again and turn to low heat. Cook for 45 minutes. Remove the heat and puree the mixture with a hand blender. Drizzle with whipping cream and sprinkle with some ground cinnamon. Serve.

GIGI NOTES

In Western cooking, mirepoix (celery, carrot and onion) creates the background flavour in many dishes. I also used it in the recipe of roast turkey in Chinese style on p.151.

Chilli Sauce with Dried Herring and Salted Olives

INGREDIENTS

4 cups oil
2 cups grated garlic
2 cups finely chopped bird's eye chillies
2 cups dried Sakura shrimps (slightly rinsed, drained and ground in a food processor)
3 cups dried silver stripe round herring (slightly rinsed and drained)
1 cup Chinese salted black olives (soaked in hot water for 5 minutes, finely chopped)

SEASONING

4 tbsp rice wine
4 tbsp light soy sauce
5 tbsp sugar

METHOD

1. Heat a wok and add ingredients in the order listed. Stir for 3 minutes over medium heat after each addition.
2. After stir-frying the dried herring for 3 minutes. Add seasoning and stir for 3 more minutes. Then add salted olives and toss for 2 minutes. Turn off the heat.
3. Let cool. Save in sterilized glass bottles.

GIGI NOTES

Before putting the chilli sauce into the glass bottles, blanch them in boiling water for a few minutes to kill all bacteria. Then let them dry completely before using.
When you put the chilli sauce into bottles, make sure there is a layer of oil above all ingredients. The oil will isolate the ingredients from being exposed to the air. Otherwise, the sauce may go stale and turn mouldy.

Willie's Garlic Sandwich Spread

INGREDIENTS

125 g cream cheese
60 g butter
1 tbsp dried basil
1 tbsp ground black pepper
1/2 tsp salt
1 whole head garlic (grated)

METHOD

Mix all ingredients together until well combined. Steam over high heat for 10 minutes. Leave it to cool. Serve.

GIGI NOTES

This is a recipe from my younger brother Willie. One day he called me in excitement and told me he invented a delicious sandwich spread. He even gave me the recipe so that I could try it myself. Of course, I made it and was instantly blown away. It's truly delicious and works best on thick toast with ham and avocado.
Before you season it with salt, it's better that you taste it first.

Preparing Dried Conches

METHOD

1. After you rinse the dried conches, soak them in cold water overnight. (There should be enough water to cover them.) Then pour the dried conches into a rice cooker together with the soaking water. Add 1 cube of rock sugar and 2 slices of ginger. Turn the rice cooker on and let it complete the programme. Then they are ready for use.
2. Let them cool down to room temperature and put them separately in Ziploc bags before placing them in a freezer. Before cooking, just take what you need for that meal out of the freezer and cook it straight without thawing.

Preparing Dried Shiitake Mushrooms

METHOD

1. Rinse 300 g of dried shiitake mushrooms. Soak in water until soft. Remove the stems and keep them for stewing soup.
2. Put the mushrooms and the water of mushrooms in a microwave safe bowl, add 2 slices of ginger, 2 sprigs of spring onion and 1/2 bar of brown cane sugar. Cover with cling wrap and pierce 2 holes on it. Microwave on high for 10 minutes.
3. Remove the mushrooms. Stir fry 2 cloves of garlic in 2 tbsps of oil in hot wok. Add mushrooms and then add 2 tbsps of dark soy sauce and the water from step 2. Switch to high heat and cook until there's not much liquid left. Add 2 tbsps of oyster sauce. Turn off the heat and let mushrooms stand until cool. You might divide them into smaller portions and put them into Ziploc bags and put them in the freezer. You can use the mushrooms straight from the freezer without thawing next time you need them.

黃淑儀
給入廚新手的菜式
Must-Learn Recipes For Novice Chefs

作者 Author
黃淑儀 Gigi Wong

編輯 Project Editor
Catherine Tam, Karen Kan

攝影 Photographer
Imagine Union

髮型 Hair Stylist
Jerry Liu

化妝 Makeup Artist
Chris Lam

美術統籌及設計 Art Direction & Design
Amelia Loh

出版者 Publisher
Forms Kitchen
香港鰂魚涌英皇道1065號 東達中心1305室
Room 1305, Eastern Centre, 1065 King's Road, Quarry Bay, Hong Kong.
電話 Tel: 2564 7511
傳真 Fax: 2565 5539
電郵 Email: info@wanlibk.com
網址 Web Site: http://www.wanlibk.com
http://www.facebook.com/wanlibk

發行者 Distributor
香港聯合書刊物流有限公司
SUP Publishing Logistics (HK) Ltd.
香港新界大埔汀麗路36號 中華商務印刷大廈3字樓
3/F., C&C Building, 36 Ting Lai Road, Tai Po, N.T., Hong Kong
電話 Tel: 2150 2100
傳真 Fax: 2407 3062
電郵 Email: info@suplogistics.com.hk

承印者 Printer
中華商務彩色印刷有限公司
C&C Offset Printing Co., Ltd.

出版日期 Publishing Date
二零一九年七月第一次印刷
First print in July 2019

Published in Hong Kong by Forms Kitchen,
a division of Wan Li Book Company Limited.
ISBN 978-962-14-7011-9

鳴謝

VICTORINOX